DISCONTINUOUS AUTOMATIC CONTROL

DISCONTINUOUS AUTOMATIC CONTROL

By IRMGARD FLÜGGE-LOTZ

1953

PRINCETON UNIVERSITY PRESS

PRINCETON, NEW JERSEY

Published, 1953, by Princeton University Press
London: Geoffrey Cumberlege, Oxford University Press
L. C. CARD 52-13156

COMPOSITION BY THE PITMAN PRESS, BATH, ENGLAND
PRINTED IN THE UNITED STATES OF AMERICA

PREFACE

In automatic control systems discontinuously working elements (on-off controls) are widely used. Their principal advantage is derived from their simple construction, but they produce complicated phenomena in the controlled system. Many unforeseen features have been discovered from studies of the transient vibrations of these systems.

The differential equations that describe motions controlled in such a manner are nonlinear, although they may appear to be linear for a limited time interval.

Nearly every publication on servomechanisms mentions discontinuous control systems. Mechanical systems having *no restoring force* other than that supplied by the discontinuous control were studied as early as 1934 (H. L. Hazen, *Journal of the Franklin Institute*, Vol. 218). Mechanical systems which *have restoring forces*, other than that of the control device, form the subject of this monograph. Their motions under the influence of discontinuous controls are studied in detail. The methods and results are not restricted to mechanical systems, but they apply also to thermodynamic and electrical systems whenever the differential equations are of the same form.

This book is the outcome of research conducted during the war and which has been continued in recent years. I am indebted to Dr. K. Klotter, now professor at Stanford University, for his participation in the earlier phase of the work, and to Mr. H. N. Abramson for his valuable assistance during two years, particularly for his help in choosing examples and preparing the final figures. Additionally I have to thank him for his help in reading the proofs. Mr. W. S. Wunch gave valuable assistance in the final polishing of the text. I also wish to express my thanks to Professor G. Bock, formerly director of the Deutsche Versuchsanstalt für Luftfahrt, Berlin-Adlershof, for the encouragement and administrative support received from him during the war, and to the Office of Air Research and the Office of Naval Research, whose financial assistance made it possible to continue the work and to write an extensive report, modified slightly in the form presented here. The list of persons I am indebted to would not be complete without mentioning Prof. S. Lefschetz, who encouraged this publication, and my husband Prof. W. Flügge, with whom I often discussed the results of my studies and their representation.

IRMGARD FLÜGGE-LOTZ

February, 1953
Stanford University

CONTENTS

DISCONTINUOUS AUTOMATIC CONTROL

INTRODUCTION

In various technical problems disturbances of the main motion of the body start undesired oscillations. If these oscillations are damped, the body may return to its former motion and in many cases even to its desired path. However, in some cases the damping of the disturbance is unsatisfactory, and in others, even missing. In these cases a special action has to occur in order to eliminate the consequences of such disturbances. They can be eliminated by observing the deviation from the desired motion and by applying forces which counteract these deviations.

As an example, one may take the motion of a missile or an airplane which is expected to fly straight ahead in air at rest. If it meets a gust, it will start undesired oscillations. If the missile is stable these will be damped. However, the amplitudes of the oscillatory motion are often so large during the essential time of action that they have to be decreased faster by special devices. Initially, one thinks of a continuously correcting device.

Continuously working control-instruments have been used for a long time. The centrifugal regulator for steam engines is a classical example. However, there are well-known types of regulators which work discontinuously. One such device is the thermostat controlling the heating of a house. It works in this way: if the actual temperature is lower than the desired temperature by a certain amount, the furnace will start heating and stop at the moment the desired temperature is reached. Instruments of this type are usually simply constructed and therefore less expensive. It appears desirable to investigate whether an analogous control could be used for missiles, since in this case all instruments are destroyed when the missile crashes, so that there is special interest for obtaining the best control at least expense.

The following treatment of the problem of discontinuous control is not restricted to special applications. The only assumptions made are that the uncontrolled motion should be described by a linear differential equation or by a system of such equations. In other words, the uncontrolled motion should be represented by the superposition of damped or undamped sinusoidal *oscillations* or by exponential *functions*.

To correct the disturbed main motion or the disturbed rest, the controlling mechanism senses the deviations from the desired state, picking up a combination of characteristic variables of the motion which describe these deviations.

The discontinuous control can then be applied in different forms:

(1) Correcting forces (or moments) of constant magnitude change their sign discontinuously under certain well-defined conditions (switch points occur at the zero-crossing of the control function).

(2) The applied corrective forces or moments are assumed to be proportional to the displacement of a particular device (e.g., rudder). This device moves with constant velocity, which changes its sign discontinuously under prescribed conditions. In each interval in which the control mechanism applies a constant force or moment or in which it produces a constant rate of change of forces or moments (between two switch points) the motion of the body continues to be governed by linear differential equations. However, the differential equations for the $(2n + 1)$th and the $2n$th intervals are different, they differ in the sign of those terms which show the influence of the control device. Since the motion of the body is supposed to be continuous, the values of the phase variables at the end of one interval determine the values of the phase variables at the start of the next one. It is obvious that the final state of motion does not depend linearly on the initial disturbance of the body; thus the motion presents a nonlinear problem in mechanics. It is the purpose of this book to give a survey of the types of motion that may occur and to study their dependence on the special type of control chosen.

The representation of the motion of the controlled system in the phase plane, or in the phase space, has proved to be very useful for those purposes. Graphical constructions could be employed primarily. Thus engineers who prefer graphical aids to complicated computational procedures can more readily obtain solutions to this nonlinear problem.

First assume the problem to be attacked with a perfectly functioning instrument; that is, assume that zero-crossing and switching coincide exactly. It is readily shown that this idealized problem does not always have a solution and that it is occasionally necessary to adapt the analytic treatment more closely to reality. In other words, one must account for the slight delay of switching in relation to the zero crossing of the control function—such as may be caused by friction and various other mechanical imperfections of the control instrument. Thus "imperfect discontinuous control" is studied in many details and found to give very satisfactory results.

Owing to the nonlinearity of the problem, three very different types of motion occur, two of which have to be avoided:

(1) *Motions which tend toward undamped periodic motions.* In general, these are not desirable. They may be desirable if the asymptotic periodic state, is a motion with very small amplitudes and high frequencies.

(2) *Motions which tend toward a motion with control at rest.* These too are undesirable. In some cases control at rest means the same damping as the uncontrolled motion, but a shifted neutral position. In others control at

rest means that one coordinate of the motion grows indefinitely, the motion becomes unstable.

(3) *Motions which become undetermined at certain well-defined points in the phase plane (or phase space) with the assumption of idealized control.* Here the study of imperfect control reveals the most interesting features of this nonlinear vibration problem. It shows that high frequency motions may suddenly occur after many periods of low frequency motion. Since this high frequency vibration may have small amplitudes, the average motion represents a highly desired one in many instances.

Some specially controlled motions will be studied in detail; thus the basic general features of discontinuously controlled motions will become apparent. Variations of the control function by the introduction of other variables, analytic functions, or nonlinear terms should not pose serious problems.

The theory is sufficiently developed to allow the design of discontinuous control systems with optimum efficiency.

PART I

DISCONTINUOUS CONTROL OF A MOVING BODY WITH A SINGLE DEGREE OF FREEDOM

1. THE DIFFERENTIAL EQUATION OF MOTION

The undisturbed motion of a body with a single degree of freedom may be characterized by $\phi_0(t)$. If this state is disturbed, the deviation ϕ may be determined by the differential equation

$$a^*\ddot{\phi} + 2b^*\dot{\phi} + c^*\phi = 0 \tag{1}$$

in which a^*, b^*, c^* are all greater than zero. Equation (1) will be called the "equation of state." The solution of this equation is given by

$$\phi = C^* e^{-\frac{b^*}{a^*}t} \cos(\omega t + \varepsilon^*) \tag{2}$$

with

$$\omega = \frac{1}{a^*}\sqrt{c^*a^* - b^{*2}} \tag{3}$$

in which C^* and ε^* are constants of integration determined by the initial conditions of motion. The various results obtainable, corresponding to the given values of a^*, b^*, and c^*, are well known.

The motion of the body, described by ϕ, may be influenced by a correcting force or moment (force if ϕ is a length measurement; moment, if ϕ determines angular displacement) in order to obtain a greater degree of damping. In this case the equation of state is

$$a^*\ddot{\phi} + 2b^*\dot{\phi} + c^*\phi = M^* = N^*\beta \tag{4}$$

where β is the force- or moment-producing element and N^* is its coefficient of efficiency (N^* is always larger than zero). As an example, consider the long-period longitudinal motion of an airplane neglecting other motions. β would represent the angle of the horizontal stabilizer used for influencing the undesired oscillations. In an automatically controlled motion the control element[1] β depends upon the coordinate ϕ and its derivatives. The equation

$$G(\beta, \dot{\beta}, \ldots) = F(\phi, \dot{\phi}, \ldots) \tag{5}$$

connecting ϕ and β is called the "equation of control" (regulator equation).

[1] Sometimes called the "manipulator."

2. DESCRIPTION OF THE CONTROLLING MECHANISM

2.1. Discontinuously applied forces or moments. The control element β is assumed to depend upon ϕ and $\dot{\phi}$ by the relationships[1]

$$\beta = -\beta_0 \operatorname{sgn} (\rho_1\phi + \rho_2\dot{\phi}) \tag{6a}$$

or

$$\beta = +\beta_0 \operatorname{sgn} (\rho_1\phi + \rho_2\dot{\phi}) \tag{6b}$$

with $\beta_0 > 0$ and $\rho_1 > 0$. In either case the control element may have two different values of the same absolute size but of opposite sign.

The function defined by

$$F = \rho_1\phi + \rho_2\dot{\phi} \tag{6c}$$

is called the "control function." The sign of this function determines in equations (6a) and (6b) the sign of the control element. The control function F in passing the value zero "commands" the control element β to change its position from $+\beta_0$ to $-\beta_0$ or from $-\beta_0$ to $+\beta_0$. The points at which the control element β changes its sign are called "switch points."

We may now write

$$F = \rho_1 \left(\phi + \frac{\rho_2}{\rho_1}\dot{\phi}\right) = \rho_1 (\phi + \rho\dot{\phi}) \tag{7}$$

with $\dfrac{\rho_2}{\rho_1} = \rho$ along with the previous assumption $\rho_1 > 0$.

Fig. 1 indicates the influence of the sign of ρ. If $\rho = 0$ the zero values of ϕ and F occur at the same instant; if $\rho > 0$ the zero value of F will precede the zero value of ϕ; if $\rho < 0$ the zero value of F will succeed the zero value of ϕ ("leading" and "lagging" control functions).

The controlling force or moment M^* (see equation (4)) is $M^* = N^*\beta$. Fig. 2 shows the moment corresponding to equations (6a) and (6b). These two equations then define two different controlling systems A and B:

$$\text{System A:} \quad M^* = N^*\beta = -N^*\beta_0 \operatorname{sgn} \rho_1(\phi + \rho\dot{\phi}) \tag{8a}$$

$$\text{System B:} \quad M^* = N^*\beta = +N^*\beta_0 \operatorname{sgn} \rho_1(\phi + \rho\dot{\phi}) \tag{8b}$$

---

[1] The "signum" symbol is defined according to the following equations:

$$\operatorname{sgn} x = \frac{x}{|x|} = +1, \quad \text{if } x > 0; \quad \operatorname{sgn} x = -1, \quad \text{if } x < 0.$$

The advantages and disadvantages of systems A and B will be thoroughly discussed later, but we may obtain a qualitative grasp of their meaning at this time.

(1) If $|\rho|$ is small, then $F \approx \phi$ and sgn $F \approx$ sgn ϕ; the equation of state will be the following (see equation (4)):

$$a^*\ddot{\phi} + 2b^*\dot{\phi} + c^*\left(\phi \pm \frac{N^*\beta_0}{c^*}\operatorname{sgn}\phi\right) = 0 \tag{9}$$

This means that in system A (upper sign in equation (9)) the restoring couple $c^*\phi$ is reinforced; in system B it is diminished.

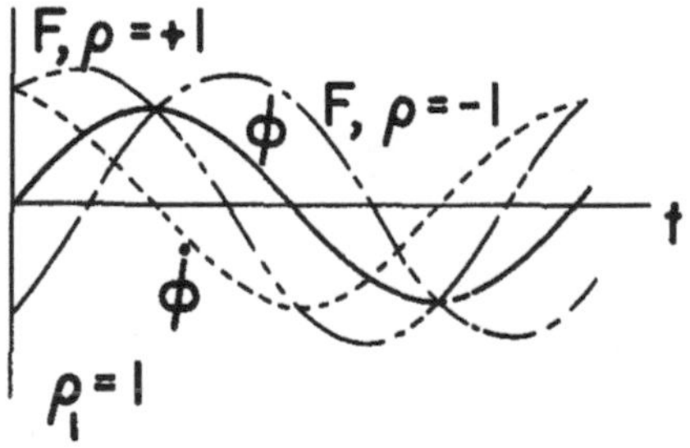

FIG. 1

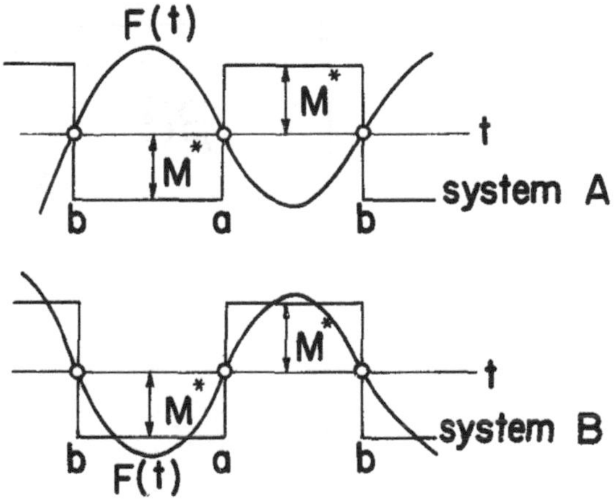

FIG. 2

(2) If ρ has a very large value, then $F \approx \rho\dot{\phi}$ and sgn F has the sign of $\dot{\phi}$. In this case we find

$$a^*\ddot{\phi} + 2b^*\left(\dot{\phi} \pm \frac{N^*\beta_0}{2b^*}\operatorname{sgn}\rho\dot{\phi}\right) + c^*\phi = 0 \tag{10}$$

so that for system A (upper sign) there is a reinforced damping.

This brief analysis implies that system A may be more favorable than system B; this idea will be confirmed later.

2.2. Discontinuous velocity of application of forces or moments. The control element β is assumed to depend upon ϕ and $\dot{\phi}$ by the following relation (corresponding to equations (6a, b) in the case of position control):

$$\frac{d\beta}{dt} = \dot{\beta} = \mp V\operatorname{sgn}F = \mp V\operatorname{sgn}(\rho_1\phi + \rho_2\dot{\phi}) \tag{11}$$

with $V > 0$ and $\rho_1 > 0$. Again there are two controlling systems which may be defined as follows:

$$\text{System } C_1: \quad \dot{\beta} = -V \operatorname{sgn}(\rho_1\phi + \rho_2\dot{\phi}) \tag{12a}$$

$$\text{System } C_2: \quad \dot{\beta} = +V \operatorname{sgn}(\rho_1\phi + \rho_2\dot{\phi}) \tag{12b}$$

However, it is not possible to conceive of the meaning of these two different systems as easily as in the case of the discontinuously applied forces or moments.

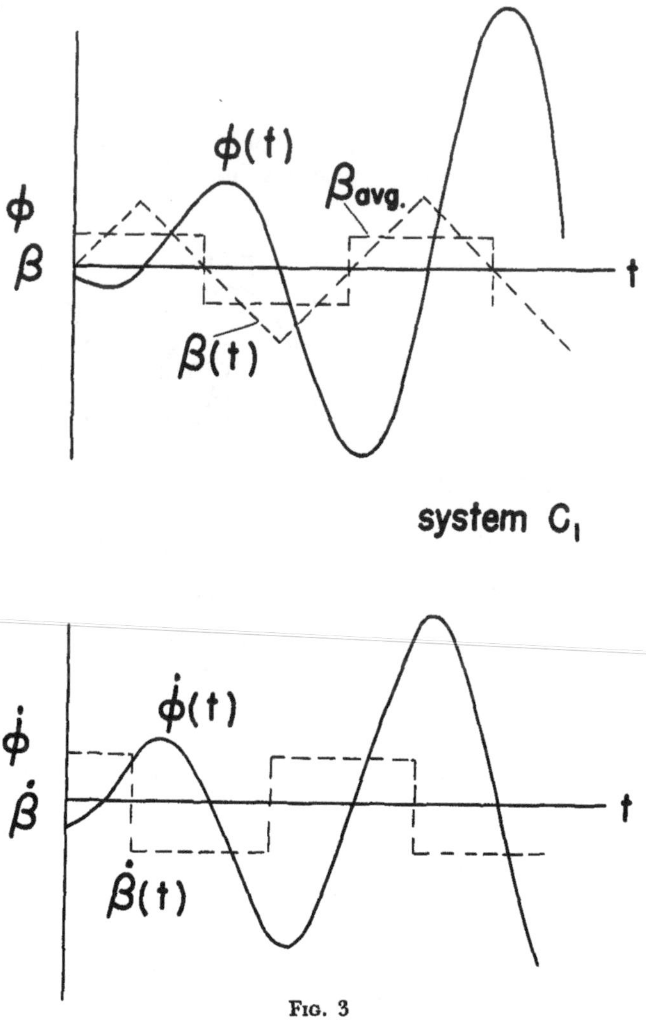

FIG. 3

The functions ϕ and β and their time derivatives $\dot{\phi}$ and $\dot{\beta}$ are shown in Fig. 3 for a system of type C_1. In the upper diagram an average value of β has been shown so that one may easily see that there is an analogy to the case of the discontinuously applied force—in this case corresponding to system A with negative ρ (Figs. 1 and 2 will aid in verifying this statement). In order to establish such an analogy it is necessary that $\beta(0) = 0$, but this fact is not always to be expected. Suppose that $\beta(0)$ is large negatively, in which case $\beta(t)$ may never attain positive values. This means that, corresponding to the size and sign of $\beta(0)$ it is possible that $\beta(t)$ never changes its sign (see Fig. 4). We conclude that the replacement of a discontinuous β-control by a discontinuous $\dot{\beta}$-control may smooth the motion under special conditions, but in general entirely different types of motion will be produced.

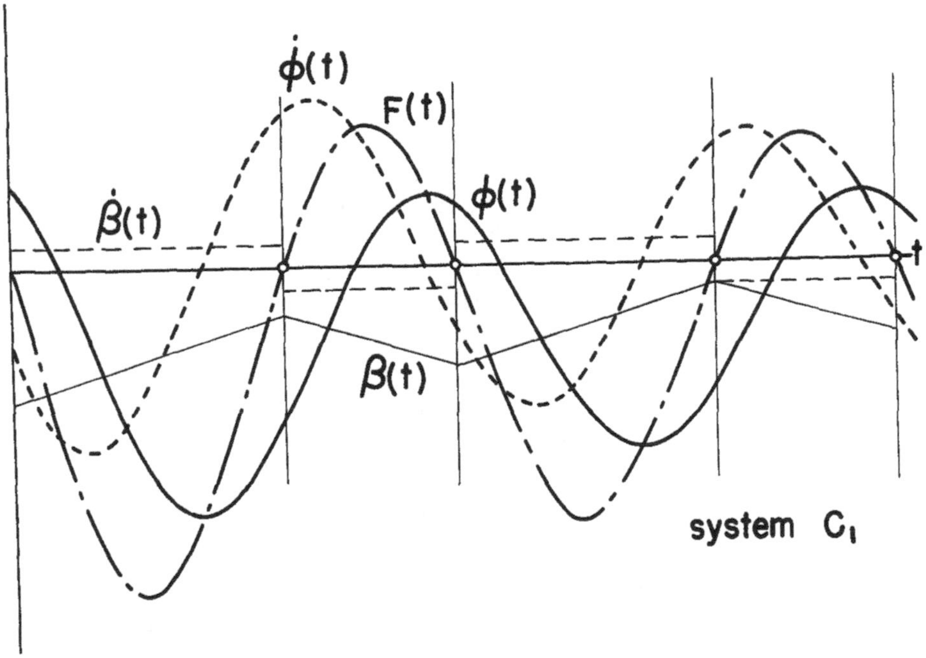

FIG. 4

The results are altered considerably if the $\dot{\beta}$-control system contains a "feedback" (follow-up). In this case the control function is

$$F = \rho_1\phi + \rho_2\dot{\phi} + \rho_3\beta$$

and we have

$$\dot{\beta} = \mp V \operatorname{sgn}(\rho_1\phi + \rho_2\dot{\phi} + \rho_3\beta) \tag{13}$$

In this type of control system it is quite possible to have a control element β which at times has a positive position and at other times a negative position. Therefore, such a system may have qualities similar to those of the position control, but it will work more smoothly.

2.3. Imperfections of the controlling mechanism. In the preceding sections the different control mechanisms are described under the assumption that they are perfect; in other words the control function is produced without error, and the control element will immediately change its position or velocity when F crosses zero. An actual mechanism will show how far these assumptions are justified.

In general a control mechanism will consist of three essential parts:

(1) The mechanism must contain an element which detects and measures the undesired deviations. In our case the element will consist of two components: one to measure ϕ and a second to measure $\dot{\phi}$. In the particular case of the control of the symmetrical motion of a guided missile (for details see Part III) ϕ is the deviation from a desired constant angle of pitch, and $\dot{\phi} = d\phi/dt$ is the rate of change of the pitch angle.

(2) A "mixing" device is required. In this instrument the values of ϕ and $\dot{\phi}$ are introduced, multiplied by the coefficients ρ_1 and ρ_2 (these are

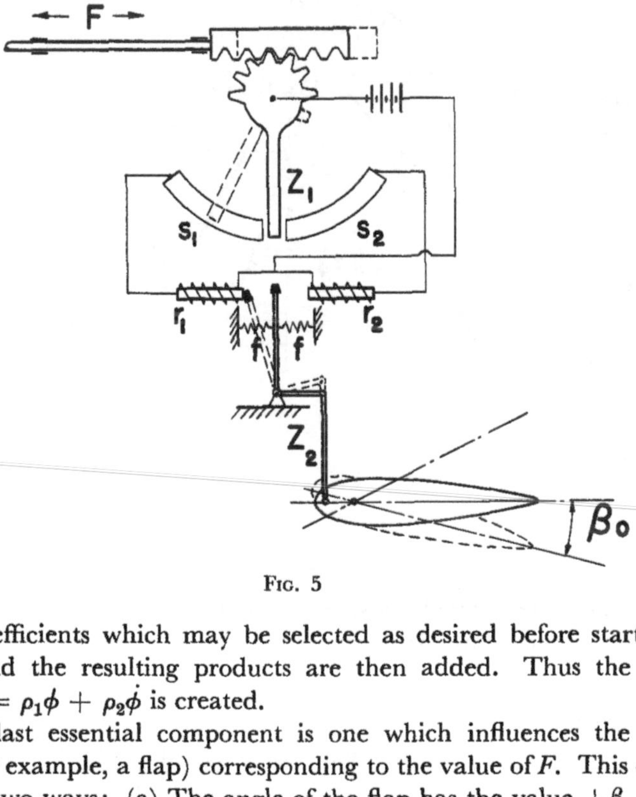

FIG. 5

constant coefficients which may be selected as desired before starting the motion), and the resulting products are then added. Thus the control function $F = \rho_1\phi + \rho_2\dot{\phi}$ is created.

(3) The last essential component is one which influences the control element (for example, a flap) corresponding to the value of F. This element may act in two ways: (a) The angle of the flap has the value $+\beta_0$ or $-\beta_0$ corresponding to the sign of the control function F. (b) The angle of the flap changes with constant angular velocity $+V$ or $-V$ corresponding to the sign of F.

Each of these three elements will be a mechanical or electrical system subject to vibrations. The deviations ϕ and $\dot{\phi}$ may be "picked-up" by gyroscopes[1] (ϕ by a bank and climb gyro, $\dot{\phi}$ by a gyroscopic turn indicator).

[1] See Eduard Fischel, Die vollautomatische Flugzeugsteuerung (The Automatic Control of an Airplane), *Ringbuch der Luftfahrttechnik*, V, E4, 1940. This paper gives detailed descriptions of the instruments in use. Some information may also be found in George E. Irvin, *Aircraft Instruments*, McGraw-Hill, New York, 1941.

These deviations ϕ and $\dot{\phi}$ may then be converted into an electric current and mixed in another electrical instrument. The mixing device could be a mechanical device using the scale-readings of the two gyroscopic instruments directly.

The mechanism which changes the position or velocity of the control element (flap) will contain a relay in order to produce discontinuities. Fig. 5 shows element (3) in the case of position control. The position of the hand Z_1 is given by the value of the control function F produced by the mixing device. The imperfections of elements (1) and (2) will influence

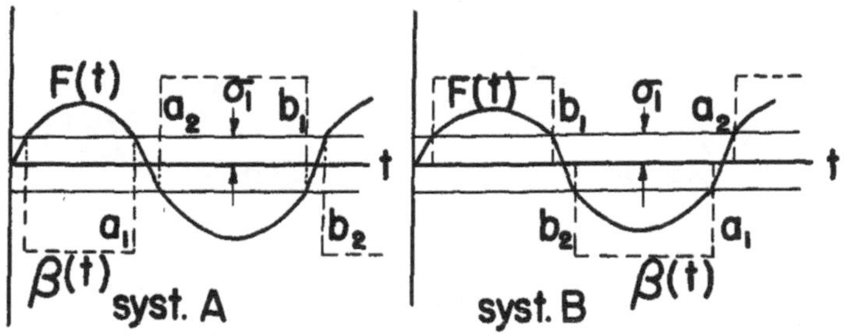

FIG. 6

the exactness of the control function F, and imperfections of element (3) will disturb the relation between F and the position, or the velocity, of the control element (flap).

In case that the control function contains β

$$F = \rho_1\phi + \rho_2\dot{\phi} + \rho_3\beta,$$

the mixing device adds to $\rho_1\phi + \rho_2\dot{\phi}$ the quantity $\rho_3\beta$ (β being "fed back"). Here imperfect creation of β modifies the control function.

Of the three sources of imperfections mentioned, only the influence of one will be studied in detail: imperfection of element (3), which disturbs the coincidence of F crossing zero and the switching. There are several different types of imperfections of element (3).

A neutral zone (or "dead" zone) is easily produced by the unavoidable insulation between the conductive parts S_1 and S_2. Fig. 6 illustrates this case. The control element β does not respond at all if $|F|$ is smaller than a certain value σ_1. This neutral zone $2\sigma_1$ has a variable "dead time."

System A		System B		
$F > \sigma_1$	$\beta = -\beta_0$	$F > \sigma_1$	$\beta = +\beta_0$	(14)
$\sigma_1 \geq F \geq -\sigma_1$	$\beta = 0$	$\sigma_1 \geq F \geq -\sigma_1$	$\beta = 0$	
$F < -\sigma_1$	$\beta = +\beta_0$	$F < \sigma_1$	$\beta = -\beta_0$	

Other imperfections may occur in the following ways. The motion of the arm Z_2 is guided by the direction of the current either through the circuit containing the coil r_1 or through that containing r_2 and is restrained by springs f. The arm Z_2 will need some time for changing from contact with r_1 to one with r_2 or vice versa. Thus the coincidence of $F = 0$ and "switching" is disturbed. Element (3) is a vibratory circuit. An exact theory should

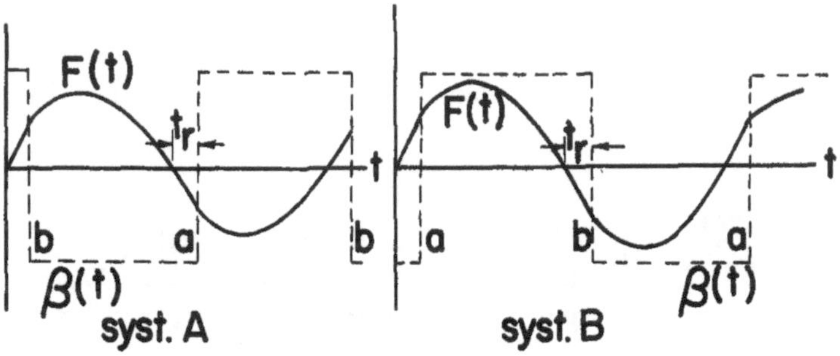

FIG. 7

use this fact to establish the connection between F and the position of the control element (flap). However, the problem would then become complicated in an undesirable manner. The delay of action of the flap may be taken into account by either (a) stating that the switching of Z_2 follows the zero crossing of F after a time t_r (constant time lag, Fig. 7 illustrates this type of

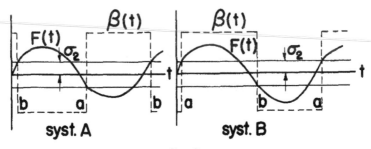

FIG. 8

imperfection); or (b) the switching occurs when the control function has attained a certain fixed absolute value σ_2 ("threshold") after passing zero (see Fig. 8):

$$\left.\begin{array}{l}\beta \text{ changes if } F = -\sigma_2 \text{ for } \dot{F} < 0 \\ \beta \text{ changes if } F = +\sigma_2 \text{ for } \dot{F} > 0.\end{array}\right\} \qquad (15)$$

Both of these assumptions may be looked upon as limiting cases of the actual imperfection. The theory will show that their effects will allow us to judge the influence of variable t_r or variable threshold σ_2.

If a velocity control mechanism is used, the position of Z_2 would not give the position of the controlling flap immediately. In this event the position of the arm Z_2 is utilized to change the direction of a motor running at constant speed. This motor changes the position of the flap β with constant speed $+V$ or $-V$ corresponding to the situation of Z_2. If the control system is expected to act for a short time only in order to fulfill its purpose, it will be more economical to use a pneumatic device to alternate the direction of a constant speed mechanism instead of a motor. The necessary energy is obtained from stored compressed air.

3. THE SOLUTION OF THE DIFFERENTIAL EQUATION OF MOTION AND ITS CONTINUITY QUALITIES

3.1. Position control. The equations describing the motion of the body for this type of control have been given as:

$$a^*\ddot{\phi} + 2b^*\dot{\phi} + c^*\phi = M^* = N^*\beta \tag{16}$$

$$\beta = \mp\beta_0 \operatorname{sgn} F, \quad \text{with } F = \rho_1\phi + \rho_2\dot{\phi} \tag{17}$$

These equations are valid only for an ideal control mechanism. Before obtaining the finite solution of this system of equations, one should attempt to reduce the number of parameters as far as possible in order to find those parameters which may be considered as essential to the problem.

Equation (16) has four parameters: a^*, b^*, c^*, and N^*. Dividing the entire equation by the inertia term a^* and setting

$$\frac{c^*}{a^*} = \omega^2; \qquad \frac{2b^*}{\sqrt{c^* \cdot a^*}} = 2D; \qquad \frac{N^*}{a^*} = N \tag{18}$$

we find

$$\frac{\ddot{\phi}}{\omega^2} + \frac{2D}{\omega}\dot{\phi} + \phi = \frac{N}{\omega^2}\beta \tag{19}$$

A new dimensionless time variable may be introduced so that $\tau = \omega t$ and $d\tau = \omega dt$. Equation (19) becomes

$$\phi'' + 2D\phi' + \phi = \frac{N}{\omega^2}\beta \tag{20}$$

where a prime indicates differentiation with respect to τ.

Equation (17) may be rewritten as

$$\beta = \mp\beta_0 \operatorname{sgn} F, \quad \text{with } F = \rho_1\phi + \omega\rho_2\phi' \tag{21}$$

For further simplification ϕ and β may be replaced by two new variables ψ and β^*:

$$\psi = \frac{\phi}{(N\beta_0/\omega^2)} \tag{22}$$

and

$$\frac{\beta}{\beta_0} = \beta^* \tag{23}$$

This substitution means that ϕ is to be measured in a special unit, $N\beta_0/\omega^2$, which characterizes the ratio of the controlling and restoring couples. Combining equation (20) and (21) and introducing the new variables we have

$$\psi'' + 2D\psi' + \psi = \mp \text{ sgn } \left[\frac{N\beta_0}{\omega^2}(\rho_1\psi + \omega\rho_2\psi')\right] \tag{24}$$

As long as we continue to restrict ourselves to an ideal mechanism, equation (24) may be simplified still further. The time at which F has the value zero depends upon the ratio $\omega\rho_2/\rho_1$. Two more substitutions may now be made.

$$\kappa = \frac{\omega\rho_2}{\rho_1} \tag{25}$$

$$F^* = \psi + \kappa\psi' \tag{26}$$

Since $\rho_1 > 0$

$$\begin{aligned}\text{sgn } F &= \text{sgn } [(N\beta_0/\omega^2)\,(\rho_1\psi + \omega\rho_2\psi')] = \text{sgn } (\psi + \kappa\psi')\\ &= \text{sgn } F^*\end{aligned}$$

and finally we have

$$\psi'' + 2D\psi' + \psi = \mp \text{ sgn } (\psi + \kappa\psi') \tag{27}$$

and

$$\beta^* = \mp \text{ sgn } (\psi + \kappa\psi')$$

Thus, for the position control problem restricted to an ideal control mechanism, we have *only two essential parameters:* D and κ.

The increment of time between two consecutive zero values of the function $F^* = \psi + \kappa\psi'$ may be termed an interval. For such an interval, the length of which is still unknown, the right side of equation (27) is constant: $+1$ or -1. Since the coefficients of the left side of equation (27) are constant, the solution for a single interval may be given quite readily:

$$\psi_m = A_{1_m}e^{\lambda_1\tau_m} + A_{2_m}e^{\lambda_2\tau_m} \mp \text{ sgn } F^*_m \tag{28}$$

for the mth interval. The coefficients λ_1 and λ_2 are

$$\left.\begin{aligned}\lambda_1 &= -D + i\sqrt{1 - D^2}\\ \lambda_2 &= -D - i\sqrt{1 - D^2}\end{aligned}\right\} \tag{29}$$

For convenience let $\tau_m = 0$ at the initial point of each interval. Then the constants A_{1_m} and A_{2_m} depend upon the initial conditions, the values of ψ_{mi} and ψ'_{mi} at $\tau_{mi} = 0$. Setting

$$A_{1,2_m} = C_m e^{\pm i\varepsilon_m}$$

we may write the solution of equation (28) in the well-known form

$$\psi_m = 2C_m e^{-D\tau_m} \cos (\nu\tau_m + \varepsilon_m) \mp \text{ sgn } F^*_m \tag{30}$$

where

$$\nu = \sqrt{1 - D^2} \tag{31}$$

If the solution for one interval is known, the solution at any time may be obtained by a step-by-step method. In other words we compute successive intervals by observing that the values of ψ and ψ' at the end of one interval are the initial values for the following interval.

In a single interval, ψ and all of its derivatives are continuous. The transition from one interval to another gives rise to a discontinuity in ψ'' of the same type as the control element β^* (equation (27)) has at the "switch points." A point where β^* changes from negative to positive values is defined as switch point a, and a point where β^* changes from positive to

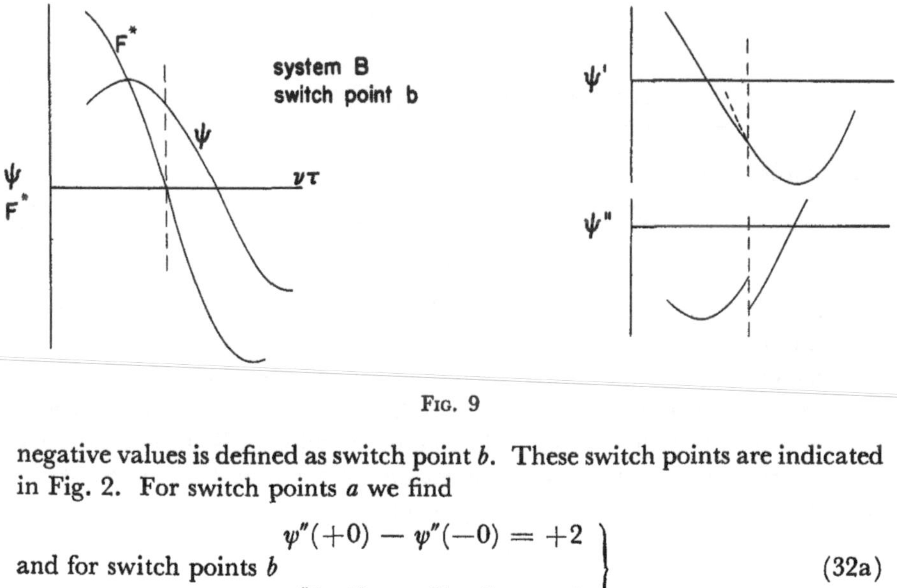

FIG. 9

negative values is defined as switch point b. These switch points are indicated in Fig. 2. For switch points a we find

$$\psi''(+0) - \psi''(-0) = +2$$

and for switch points b

$$\left.\begin{aligned} \\ \psi''(+0) - \psi''(-0) = -2 \end{aligned}\right\} \qquad (32a)$$

Therefore, the first derivative ψ' will have a discontinuous slope at those points of transition, but it is a continuous function everywhere. ψ is continuous everywhere, but it has a discontinuous curvature at the switch points. These properties are shown in Fig. 9. Furthermore, the control function $F^* = \psi + \kappa\psi'$ is continuous everywhere, but has a discontinuous slope at its zero points (switch points). Since $F^{*\prime} = \psi' + \kappa\psi''$, we see that for

$$\left.\begin{aligned} &\text{switch point } a\text{:} \quad F^{*\prime}(+0) - F^{*\prime}(-0) = 2\kappa \\ &\text{switch point } b\text{:} \quad F^{*\prime}(+0) - F^{*\prime}(-0) = -2\kappa \end{aligned}\right\} \qquad (32b)$$

These continuity qualities of the function ψ will be discussed very thoroughly later, inasmuch as they may be the cause of several very different types of motion of the oscillating body.

3.2. Velocity control. The equations describing the motion of the body for this type of control have already been given.

$$a^*\ddot{\phi} + 2b^*\dot{\phi} + c^*\phi = M^* = N^*\beta \tag{33}$$

$$\dot{\beta} = \mp V \operatorname{sgn}(\rho_1\phi + \rho_2\dot{\phi} + \rho_3\beta) \tag{34}$$

These equations are valid only for an ideal mechanism. We may treat the control either with or without feedback (no feedback implies $\rho_3 = 0$).

Let us try to find the essential parameters of the problem as we did in the case of the position control. Dividing equation (33) by the inertia term a^* and introducing the new time variable τ, we find that (compare with equation (20))

$$\phi'' + 2D\phi' + \phi = \frac{N}{\omega^2}\beta \tag{35a}$$

$$\beta' = \mp \frac{V}{\omega} \operatorname{sgn}(\rho_1\phi + \omega\rho_2\phi' + \rho_3\beta) \tag{35b}$$

Let us introduce the new variables

$$\bar{\psi} = \frac{\phi}{(N/\omega^2)(V/\omega)} \tag{36}$$

and

$$\bar{\beta} = \beta/(V/\omega) \tag{37}$$

Upon differentiating equation (35a) and introducing equation (35b), we obtain

$$\bar{\psi}''' + 2D\bar{\psi}'' + \bar{\psi}' = \mp \operatorname{sgn} F \tag{38}$$

where

$$F = (NV/\omega^3)[\rho_1\bar{\psi} + \omega\rho_2\bar{\psi}' + (V/\omega)\rho_3\bar{\beta}] \tag{39}$$

For an ideal control mechanism in which only the zero value of F is essential, F may be replaced by

$$F_1^* = \bar{\psi} + \kappa\bar{\psi}' + \mu\bar{\beta} \tag{40}$$

where

$$\kappa = \frac{\omega\rho_2}{\rho_1} \quad \text{and} \quad \mu = \frac{(V/\omega)\rho_3}{\rho_1} \tag{41}$$

ρ_1 is assumed to be positive (see p. 11). The motion of an oscillating body with one degree of freedom having discontinuous velocity control and feedback has *three essential parameters:* D, κ, and μ.

In order to determine the solution of equation (38) in a single interval, it will be found best to return to the system of equations from which equation (38) was derived:

$$\left.\begin{aligned} \bar{\psi}'' + 2D\bar{\psi}' + \bar{\psi} &= \bar{\beta} \\ \bar{\beta}' &= \mp \operatorname{sgn} F_1^* \end{aligned}\right\} \tag{42}$$

Direct integration yields

$$\bar{\beta}_m = \bar{\beta}_{mi} \mp (\operatorname{sgn} F_1^*)\tau_m \tag{43}$$

so that

$$\bar{\psi}_m'' + 2D\bar{\psi}_m' + \bar{\psi}_m = \bar{\beta}_{mi} \mp (\operatorname{sgn} F_1^*)\tau_m \tag{44}$$

The solution of equation (44) is given by

$$\begin{aligned}\bar{\psi}_m = {} & 2C_m e^{-D\tau_m} \cos(\nu\tau_m + \varepsilon_m) \\ & + [\bar{\beta}_{mi} \pm 2D \operatorname{sgn} F_1^*] \mp (\operatorname{sgn} F_1^*)\tau_m \end{aligned} \tag{45}$$

when ν is given by equation (31).

The constants C_m and ε_m of equation (45) are determined by the initial values of the mth interval: $\bar{\psi}_{mi}$, $\bar{\psi}_{mi}'$ and $\bar{\beta}_{mi}$

$$\left.\begin{aligned}\bar{\psi}_{mi} &= 2C_m \cos \varepsilon_m + [\bar{\beta}_{mi} \pm 2D \operatorname{sgn} F_1^*] \\ \bar{\psi}_{mi}' &= 2C_m \cos(\varepsilon_m + \sigma) \mp (\operatorname{sgn} F_1^*)\end{aligned}\right\} \tag{46}$$

where

$$\cos \sigma = -D \qquad \sin \sigma = \sqrt{1 - D^2} \tag{47}$$

The final values $\bar{\psi}_{me}$, $\bar{\psi}_{me}'$, and $\bar{\beta}_{me}$ of one interval are the initial values of the succeeding interval. These facts mean that these functions are continuous in the whole. From equation (38) we see that $\bar{\psi}'''$ is discontinuous at the switch points. Therefore $\bar{\psi}''$ has a discontinuous slope at these switch points and $\bar{\psi}'$ a discontinuous curvature.

With feedback the control function is $F_1^* = \bar{\psi} + \kappa\bar{\psi}' + \mu\bar{\beta}$, and without feedback it is $F_1^* = \bar{\psi} + \kappa\bar{\psi}'$. It is essential that this distinction be noted because with feedback F_1^* has a discontinuous slope at its switch points, but without feedback F_1^* is a continuous function with a continuous derivative in the whole. This fact will establish the very different qualities of these two types of velocity control.

4. COMPLETE THEORY OF THE MOTION WITH POSITION CONTROL

4.1. The representation of the motion in the phase plane.[1] The motion of an oscillating body with proportional-position control is governed by equations (27)

$$\left.\begin{aligned} \psi'' + 2D\psi' + \psi &= \beta^* \\ \beta^* &= \mp \operatorname{sgn} F^* \\ F^* &= \psi + \kappa\psi' \end{aligned}\right\} \tag{48}$$

or

$$\psi'' + 2D\psi' + \psi = \mp \operatorname{sgn}(\psi + \kappa\psi') \tag{49}$$

In a single interval (between two consecutive switchpoints) the solution of this motion is given by

$$\psi_m = 2C_m e^{-D\tau_m} \cos(\nu\tau_m + \varepsilon_m) \mp \operatorname{sgn} F^* \tag{50}$$

where $\nu = \sqrt{1 - D^2}$. For further investigations the explicit terms of the first derivative of ψ and of the control function F^* will be needed. ψ' may be written immediately as

$$\psi'_m = 2C_m e^{-D\tau_m} \cos(\nu\tau_m + \varepsilon_m + \sigma) \tag{51}$$

where

$$\cos\sigma = -D \tag{47}$$

$$\sin\sigma = \nu = \sqrt{1 - D^2} \tag{52}$$

Then from equations (48), (50), and (51)

$$F^*_m = 2C_m\sqrt{1 - 2\kappa D + \kappa^2}[e^{-D\tau_m} \cos(\nu\tau_m + \varepsilon_m + \eta)] \mp \operatorname{sgn} F^* \tag{53}$$

where

$$\eta = \operatorname{arctg} \frac{\kappa \sin\sigma}{1 + \kappa\cos\sigma} = \operatorname{arctg} \frac{\kappa\sqrt{1-D^2}}{1 - \kappa D} \tag{54}$$

The derivative of the control function F^*_m becomes

$$F^{*\prime}_m = 2C_m\sqrt{1 - 2\kappa D + \kappa^2}[e^{-D\tau_m} \cos(\nu\tau_m + \varepsilon_m + \eta + \sigma)] \tag{55}$$

[1] I. Flügge-Lotz and K. Klotter, Über Bewegungen eines Schwingers unter dem Einfluss von Schwarz-Weiss-Steuerungen, *Zentrale fur wissenschaftliches Berichtswessen der Luftfahrtforschung des Generalluftzeugmeisters* (*ZWB*), Untersuchungen und Mitteilungen Nr. 1326, Berlin, August 1, 1943; and *Zeitschrift für Angewandte Math. u. Mech.* 28 (1948) 317. Translated in *N.A.C.A. Technical Memorandum* 1237: On the Motions of an Oscillating System Under the Influence of Flip-Flop Controls, November, 1949.

The constants C_m and ε_m are determined by the initial conditions:

$$\psi_m(0) = \psi_{mi}$$
$$\psi'_m(0) = \psi'_{mi}$$

or from equations (50) and (51)

$$\begin{aligned} \psi_{mi} &= 2C_m \cos \varepsilon_m \mp \operatorname{sgn} F^* \\ \psi'_{mi} &= 2C_m \cos (\varepsilon_m + \sigma) \end{aligned} \tag{56}$$

Now that all of these relations have been established, we should be able to construct the solution $\psi(t)$ of equation (49) in a step-by-step manner, always careful to observe the sign of $F^* = \psi + \kappa\psi'$. It will prove to be more

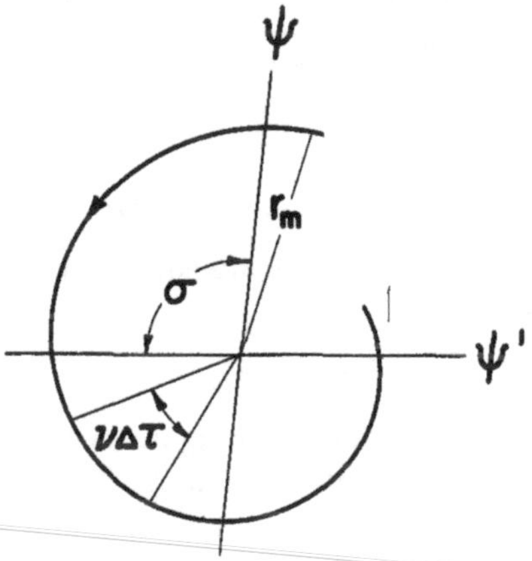

FIG. 10

convenient to construct $\psi(\psi')$ rather than $\psi(\tau)$, obtaining in this manner the "phase curve."

For an uncontrolled system the motion would be governed by the equations

$$\psi_m = 2C_m e^{-D\tau_m} \cos (\nu\tau_m + \varepsilon_m) \tag{57a}$$

and

$$\psi'_m = 2C_m e^{-D\tau_m} \cos (\nu\tau_m + \varepsilon_m + \sigma) \tag{57b}$$

Then in a phase plane having oblique axes as shown in Fig. 10, we have

$$\begin{aligned} r_m^2 &= \psi_m^2 + \psi_m'^2 - 2\psi_m\psi'_m \cos \sigma \\ &= 4C_m^2 e^{-2D\tau_m}[\cos^2(\nu\tau_m + \varepsilon_m) + \cos^2(\nu\tau_m + \varepsilon_m + \sigma) \\ &\qquad - 2 \cos (\nu\tau_m + \varepsilon_m) \cos (\nu\tau_m + \varepsilon_m + \sigma) \cos \sigma] \end{aligned}$$

By using the relation

$$\begin{aligned} \cos (\nu\tau_m + \varepsilon_m + \sigma) &= \cos (\nu\tau_m + \varepsilon_m) \cos \sigma \\ &\quad - \sin (\nu\tau_m + \varepsilon_m) \sin \sigma \end{aligned}$$

there results

$$r_m^2 = 4C_m^2 e^{-2D\tau} \sin^2\sigma$$

or

$$r_m = (2C_m \sin \sigma)e^{-\frac{D}{\nu}\nu\tau_m} \tag{58}$$

This result shows that in the phase plane having oblique axes the phase curve is a logarithmic spiral. The point of convergence of the spiral is the origin of the system. The angle between two radii from the origin to two different points of the spiral is equal to $\Delta\tau_m\sqrt{1 - D^2}$. Thus this angle is a direct measure of the time needed for this motion of the body.

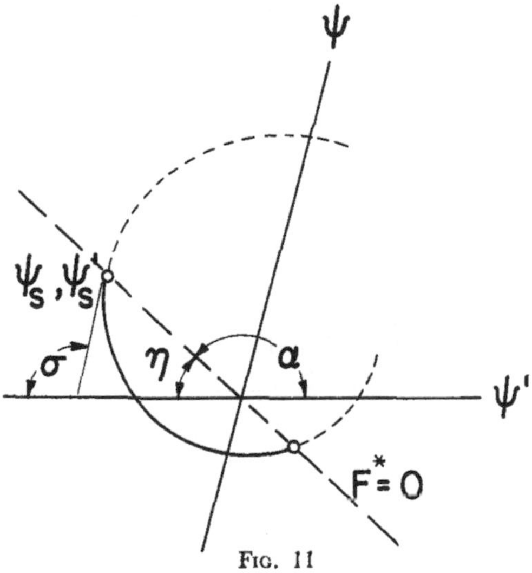

FIG. 11

For a controlled system equations (50) and (51) are used instead of equations (57a, b). Here the point of convergence of the spiral does not coincide with the origin of the system but with the point ($-$ sgn F^*) for system A and with the point ($+$ sgn F^*) for system B. Now this phase curve consists of segments of spirals of the same form, but with different points of convergence (i.e. different centers). These segments are connected at those points where F^* changes its sign.

The points ψ_s, ψ_s', for which

$$F^* = \psi_s + \kappa\psi_s' = 0 \tag{59}$$

lie on a straight line in the phase plane. This line forms an angle α with the positive direction of ψ' as shown in Fig. 11. This angle is given by

$$\begin{aligned} \tan \alpha &= \frac{\psi_s \sin \sigma}{\psi_s' - \psi_s \cos \sigma} \\ &= -\frac{\kappa \sin \sigma}{1 + \kappa \cos \sigma} \end{aligned} \tag{60}$$

From equation (54) it may be seen that

$$\tan \alpha = -\tan \eta$$

or we may conclude

$$\alpha = \pi - \eta \tag{61}$$

It will prove useful to know that the distance y_1 of a point P from the line $F^* = 0$ is proportional to the value of the control function F^* at this point. This fact is readily proved by reference to Fig. 12.

$$\begin{aligned} y_1 &= y_2 \sin(\sigma - \eta) \\ &= \psi \sin(\sigma - \eta) - \psi' \sin \eta \end{aligned}$$

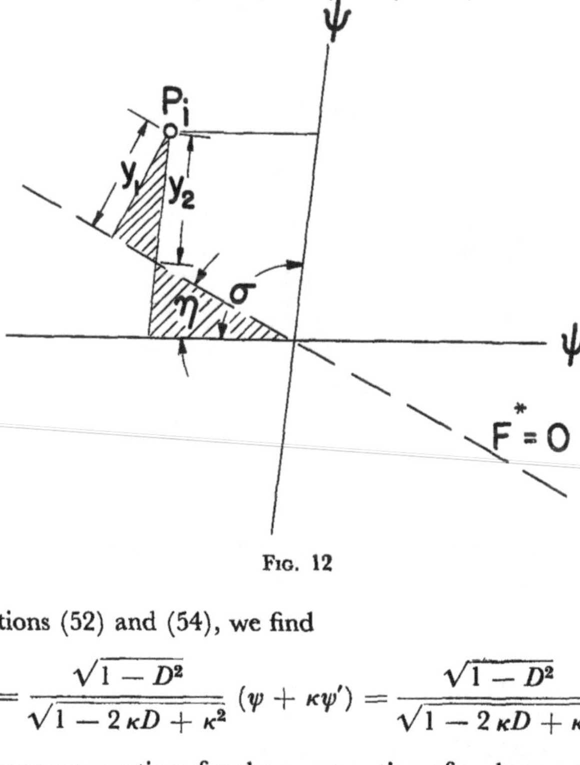

FIG. 12

Using equations (52) and (54), we find

$$y_1 = \frac{\sqrt{1 - D^2}}{\sqrt{1 - 2\kappa D + \kappa^2}} (\psi + \kappa\psi') = \frac{\sqrt{1 - D^2}}{\sqrt{1 - 2\kappa D + \kappa^2}} F^* \tag{62}$$

All necessary preparations for the construction of a phase curve have now been completed. Therefore we may focus our attention upon Fig. 13, in which the phase curve for a motion of system A having positive κ is shown. The initial values of ψ_i and ψ_i' are given. From the starting point P_i to the first switch point S_1 (first change of sign of F^*) a logarithmic spiral with its center at the point $\psi = -1$, $\psi' = 0$ is constructed. This center point will be designated as M_2. From switch point S_1 to switch point S_2 a spiral is constructed with its center at the point $\psi = +1$, $\psi' = 0$, designated M_1. From S_2 to S_3 another portion of a logarithmic spiral around M_2 is constructed. The method may easily be continued.

To construct phase curves rapidly and simply, the following procedure will prove beneficial:

(1) Prepare a logarithmic spiral (see Appendix II) of several convolutions for the given value of the parameter D.

(2) Construct the oblique system of axes on transparent paper noting the convergence points M_1 and M_2.

(3) Place the transparent paper over the spiral so that the appropriate point, M_1 or M_2 coincides with the center of convergence of the spiral.

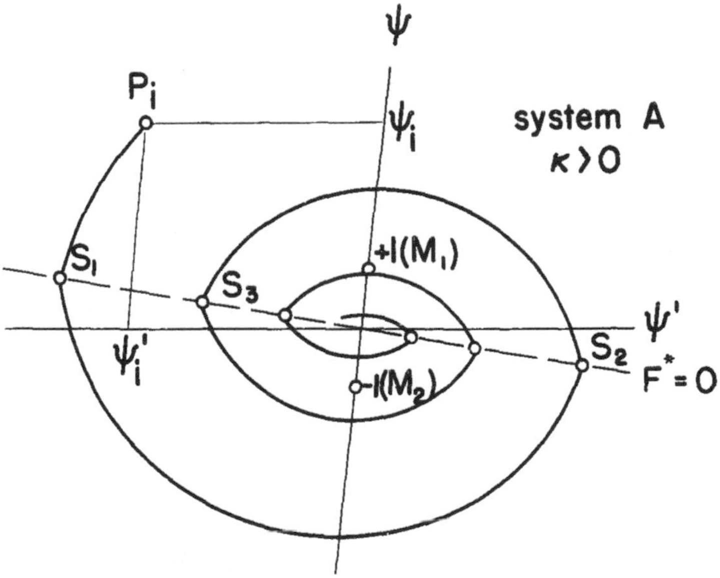

FIG. 13

(4) Now, holding these points in coincidence, shift the transparent paper so that the spiral passes through the initial point, or through the end point of the interval previously considered (this latter point lies on the line $F^* = 0$). The appropriate portion of the spiral may then be traced directly upon the transparent paper.

This represensation of the body motion by use of phase curves will greatly facilitate the study which is to follow; therefore it will be employed extensively.

4.2. Different types of motion. After relatively few examples, the construction of phase curves for systems A and B with various values for ψ_i and ψ_i' will reveal some special qualities of the body motion. Already we have seen one motion corresponding to system A in Fig. 13. Now consider the motion given in Fig. 14a for control system B. The essential difference between these two motions is obvious and is easily described. The line given by

$$F^* = \psi + \kappa\psi' = 0$$

divides the phase plane ψ, ψ' into two half planes, one corresponding to $F^* > 0$ and the other to $F^* < 0$. In system A the spirals used in one half plane have their centers in the other; in system B the spirals used in one half plane have their centers in the same half plane.

Now suppose that we are using system B. Let the configuration be such that the last switch point is so near the origin ($\psi = 0$, $\psi' = 0$) that the

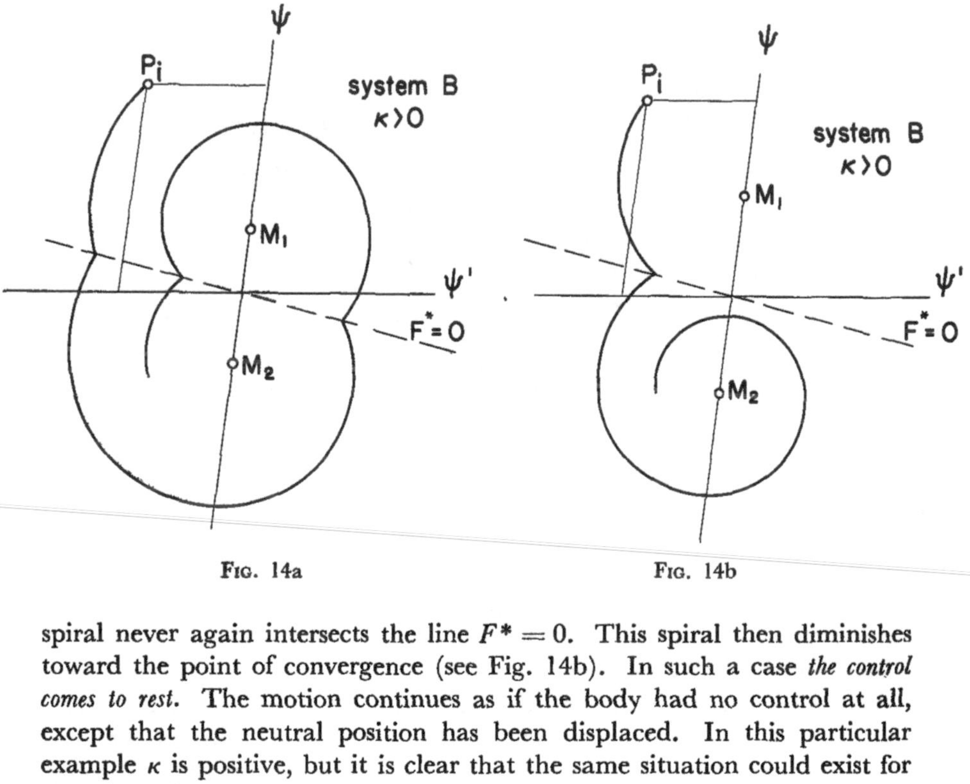

FIG. 14a FIG. 14b

spiral never again intersects the line $F^* = 0$. This spiral then diminishes toward the point of convergence (see Fig. 14b). In such a case *the control comes to rest.* The motion continues as if the body had no control at all, except that the neutral position has been displaced. In this particular example κ is positive, but it is clear that the same situation could exist for negative values of κ.

Such a "rest of control" is not possible in system A. However, system A has its own peculiarity which is apparent from Fig. 15. It is possible that in the neighborhood of the origin ($\psi = 0$, $\psi' = 0$) two consecutive switch points do not lie on opposite sides of the origin along the line $F^* = 0$. At the point S_3 of Fig. 15 the control function F^* is zero. The possibility of a continuation of the motion has yet to be discussed. It would be impossible for the motion to follow the dashed path, because it is necessary that F^* change its sign by crossing the zero value. This event cannot occur. Therefore we may conclude that for an *ideal* control system motion beyond a point possessing the qualities of S_3 in Fig. 15 is undefined.

As was briefly discussed earlier, a mechanical control system will always have some time lag inherent in its operation. Therefore β will switch some time after F^* has changed sign. If a constant time lag is assumed, we would obtain a motion such as that shown in Fig. 16. This motion has very high frequency, and it tends in the average toward the origin ($\psi = 0$, $\psi' = 0$). This average motion is determined by the equation

$$F^* = \psi + \kappa\psi' = 0$$

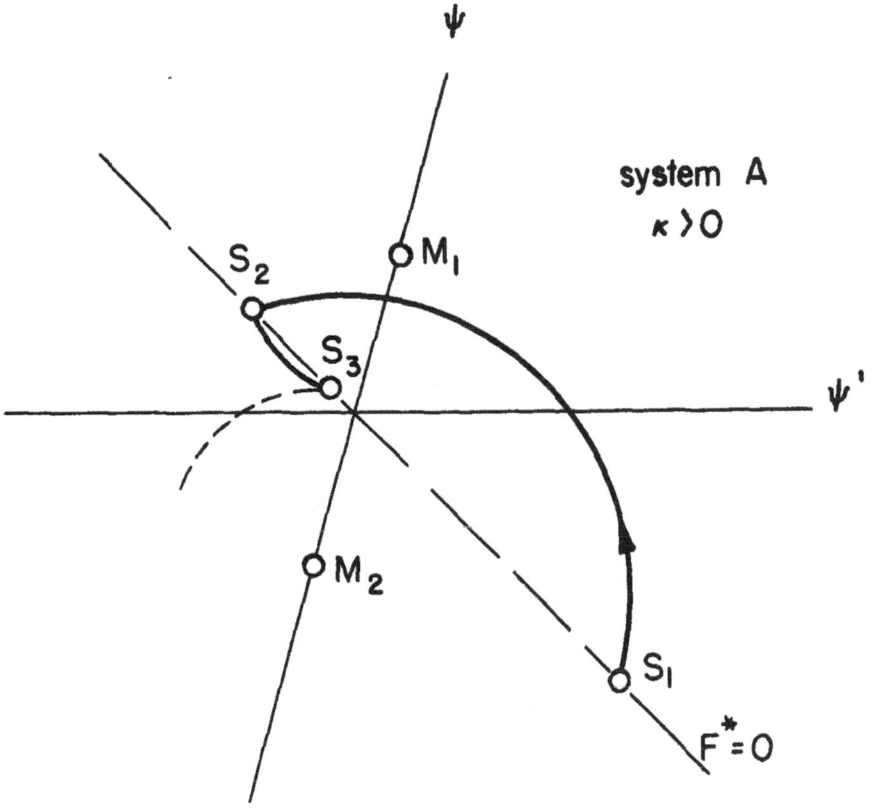

FIG. 15

The solution of this equation is given by

$$\psi = Ge^{-\frac{1}{\kappa}\tau}, \quad G = \text{constant} \tag{63}$$

This equation may be interpreted as follows: if κ is positive and small, ψ will quickly tend toward zero, or in other words the body will return to its desired path rapidly.

The occurrence of periodic motions is yet to be discussed, but they do exist. A phase curve construction for system A with a negative value of κ (as

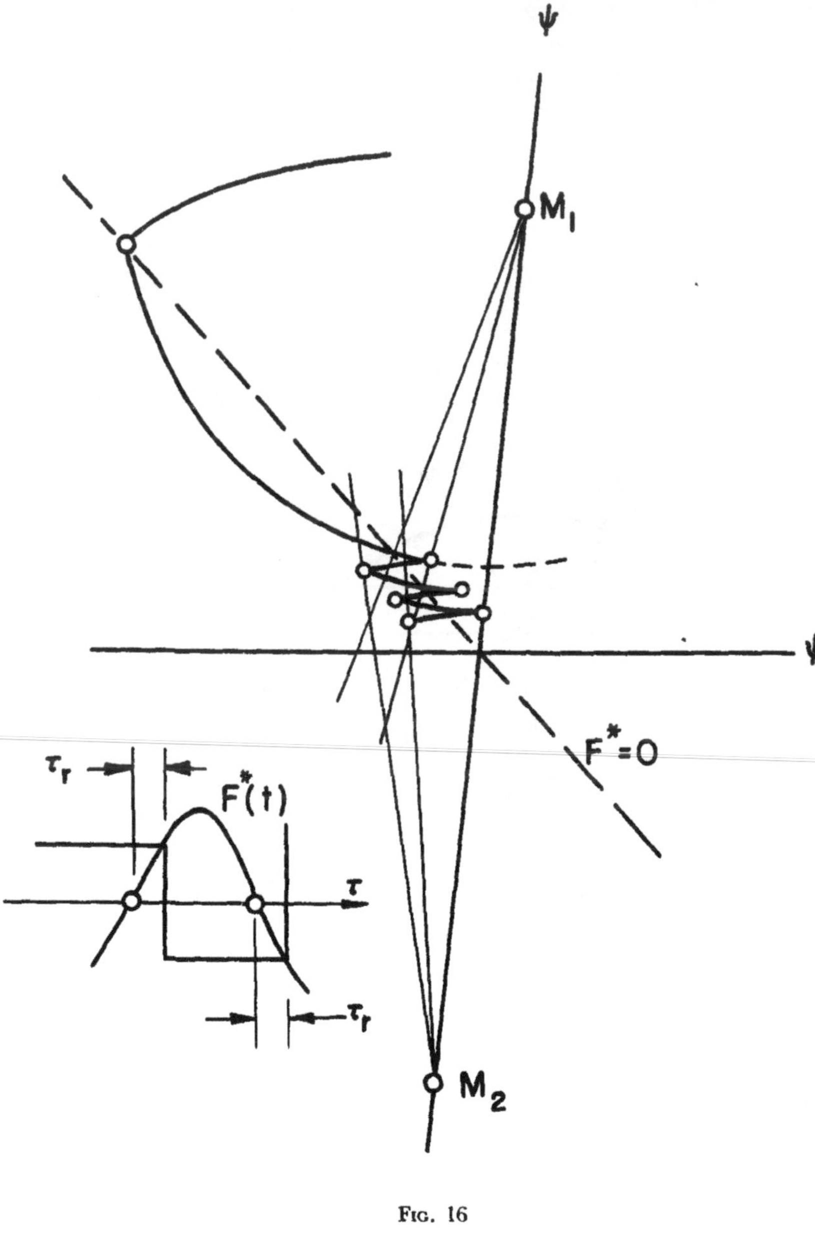

FIG. 16

shown in Fig. 17) or for system B with positive κ will lead to periodic motions. An examination of the geometry of the spirals together with the location of their centers in relation to the switching line, $F^* = \psi + \kappa\psi' = 0$, reveals that such periodic motions are only possible for system A, $\kappa < 0$ or for system B, $\kappa > 0$. A complete study of the periodic phenomena for position control (including both systems A and B) will be found in Sec. 4.5.

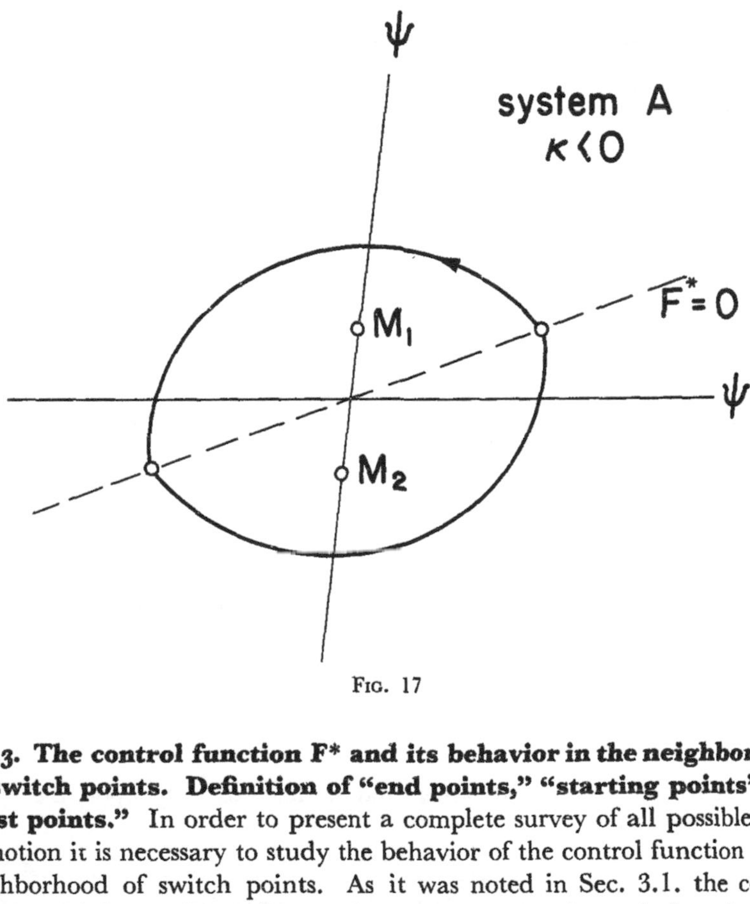

FIG. 17

4.3. The control function F* and its behavior in the neighborhood of switch points. Definition of "end points," "starting points" and "rest points." In order to present a complete survey of all possible types of motion it is necessary to study the behavior of the control function in the neighborhood of switch points. As it was noted in Sec. 3.1. the control function F^* has a discontinuous slope at every regular switch point, that is, at every point where F^* changes sign in passing through it. The importance of this statement will become more apparent later.

Fig. 18 shows the control function in the immediate neighborhood of a switch point for systems A and B with switching types *a* and *b* (see p. 20 and Fig. 2). There are four characteristic forms of $F^*(\tau)$ in the neighborhood of a switch point; all of them are shown in Fig. 18. In each sketch a regular function $F^*(\tau)$ is traced by a thin line. The discontinuity of the slope of $F^*(\tau)$ in the switch points (determined by equation 32b) is best seen by viewing the angle which $F^*(\tau)$ forms with the perpendicular to the τ axis in the switch point. There are two possibilities: the angle enlarges or diminishes in crossing the τ axis. Table I summarizes the behavior of

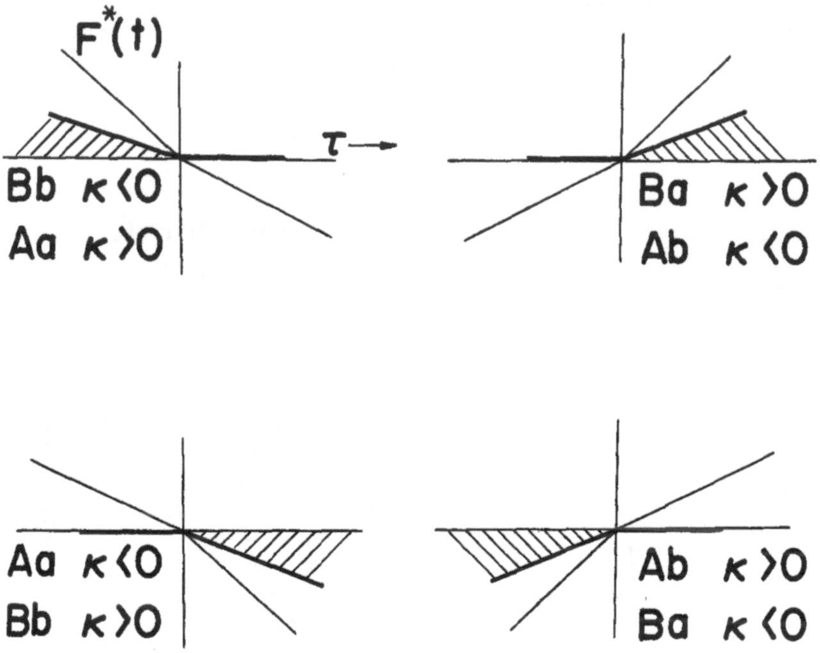

$$\beta^* = -\ \text{sgn}\ F^* \quad \text{in} \quad A$$
$$\beta^* = +\ \text{sgn}\ F^* \quad \text{in} \quad B$$

FIG. 18

$F^{*\prime}(\tau)$ in switch points, indicating that it depends only upon the system and the sign of κ.

Table I

	System A	System B
$\kappa > 0$	breaking away from the perpendicular	breaking towards the perpendicular
$\kappa < 0$	breaking towards the perpendicular	breaking away from the perpendicular

Observe now the sketch in the lower right of Fig. 18. The zero slope for $\tau > 0$ (take $\tau = 0$ as the switch point) corresponds for $\tau < 0$ to a certain

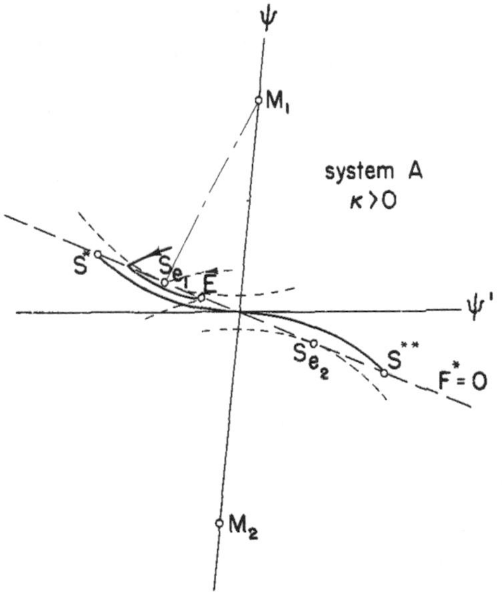

FIG. 19a

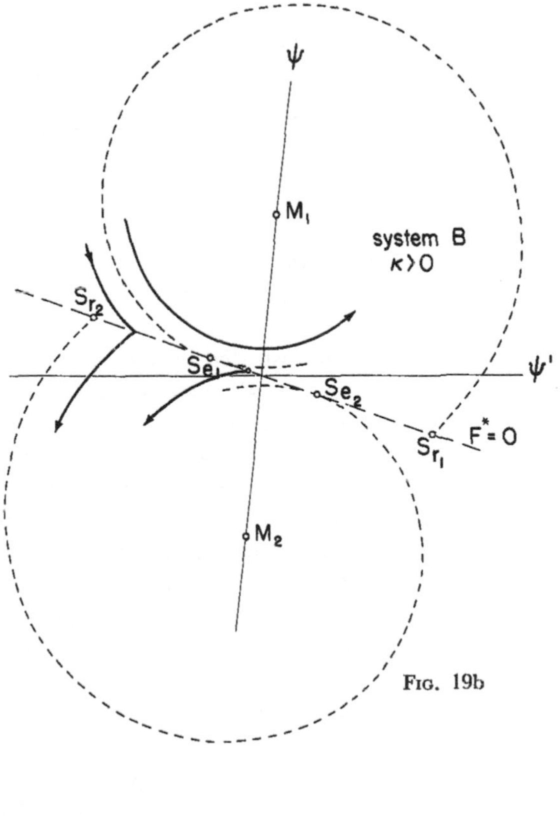

FIG. 19b

slope shown by a heavy line. All tangents lying within the shaded angular space have no corresponding values for $\tau > 0$, because a slope obeying the law of discontinuity for $\tau = 0$ would require a negative slope for $\tau > 0$. This fact is contradictory to the condition that F^* must change its sign at $\tau = 0$. Therefore, if F^* (for $\tau < 0$) lies in the shaded sector, the point $\tau = 0$ is the last point at which the motion is regular. Such points are defined as "end points." The point S_3 in Fig. 15 is such an end point.

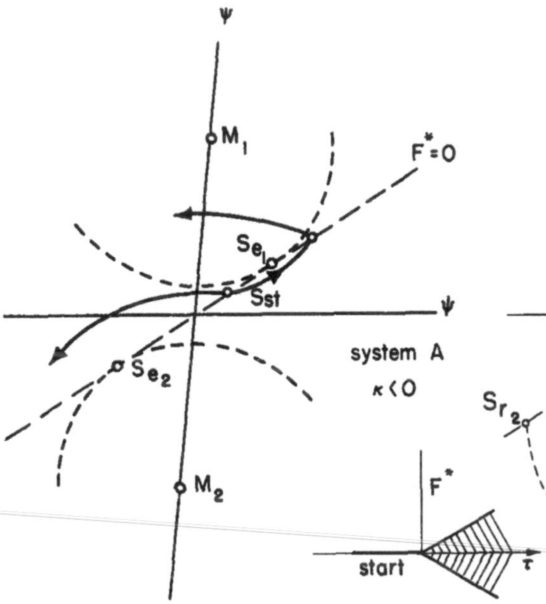

FIG. 20a

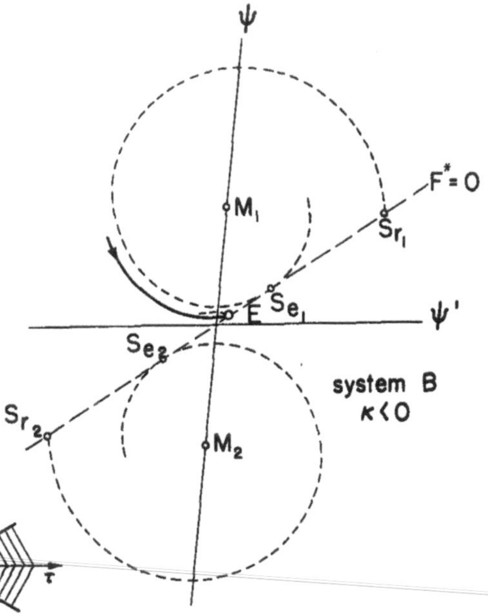

FIG. 20b

In the case just treated we have $F^{*\prime}(-0) > 0$. An analogous situation for $F^{*\prime}(-0) < 0$ is shown in the upper left of Fig. 18. There, too, end points may occur.

Another interesting case is demonstrated by the sketch in the upper right of Fig. 18. Here may be found certain values for the slope of F^* when $\tau > 0$ for which there are no corresponding values when $\tau < 0$. This indicates that certain points exist on the switching line $F^* = 0$ which may be "starting points" for a motion, but these may never be transition points. The sketch in the lower left corner indicates a starting point for $F^{*\prime}(+0) < 0$.

Figs. 19 and 20 illustrate these concepts for an actual motion. Corresponding to a system A, $\kappa > 0$, a phase curve is shown in Fig. 19a which tends toward an end point E. At the points Se_1 and Se_2 the spirals around M_1 and M_2, respectively, are tangent to the switch line, which means that at these points the control function F^* has zero slope ($F^{*\prime}(+0) = 0$; see Fig. 18Aa and Ab). All points lying on the switch line between Se_1 and Se_2 must then be end points. Likewise, all points lying on the switch line between S^*Se_1, and $S^{**}Se_2$ must start a direct motion towards an end point (S^* and S^{**} lie on the spirals which pass through the origin).

If the system of Fig. 19a is changed to one of type B, then as seen in Fig. 19b, points Se_1 and Se_2 designate those at which the control mechanism starts to rest. It is evident that all phase curves around M_1, which pass through points lying on the switch line between 0 and Sr_1, will characterize motions with control at rest. The same holds for spirals around M_2 which pass between 0 and Sr_2. The points between Se_1 and Se_2 may be starting points of motion, but they may never be transition points.

Fig. 20a shows that for system A with negative values of κ all points on the switch line between Se_1 and Se_2 are starting points of motion. Note that it is possible to start the motion at a point S_{st} with a spiral around M_2 or M_1 having a switch point of type *a* or *b*. Thus we observe that F^* may become either positive or negative starting from $F^* = 0$ for $\tau \leq 0$.

Fig. 20b demonstrates that for system B with negative κ the points on the switch line between Se_1 and Se_2 are end points for an ideal control mechanism. All phase curves which are portions of spirals passing through points between Se_1 and Sr_1 and between Se_2 and Sr_2 will characterize motions with the control at rest.

4.4. The loci of the end, starting, and rest points. In order to provide a complete theory of the controlled motion the loci of the points Se and Sr for variable κ should be determined.

The length of Se_10 for every value of κ is readily obtained. Referring to Fig. 21, we observe that the phase curve is a logarithmic spiral around M_1 given by

$$r = Ce^{-\left(\frac{D}{\nu}\right)\nu\tau} = Ce^{-[\cot(\pi - \sigma)]\nu\tau} \tag{64}$$

Therefore the angle M_1Se_10 is

$$\sphericalangle\, M_1Se_10 = \pi - \sigma \tag{65}$$

Since the angle between the switch line $F^* = 0$ and the negative ψ' axis is η, the angle at M_1 is also η. Therefore with $\overline{M_10} = 1$

$$Se_10 = \frac{\sin\eta}{\sin(\pi - \sigma)} \tag{66a}$$

Using equation (54), we find

$$Se_10 = \frac{\kappa}{\sqrt{1 - 2\kappa D + \kappa^2}} \tag{66b}$$

The value of ψ' at the point Se_1 is

$$\psi'_{e_1} = -\, Se_10\, \frac{\sin(\sigma - \eta)}{\sin\sigma} \tag{67a}$$

or

$$\psi'_{e_1} = \frac{-\kappa}{1 - 2\kappa D + \kappa^2} \tag{67b}$$

since

$$\sin(\sigma - \eta) = \frac{\sqrt{1 - D^2}}{\sqrt{1 - 2\kappa D + \kappa^2}} \tag{68}$$

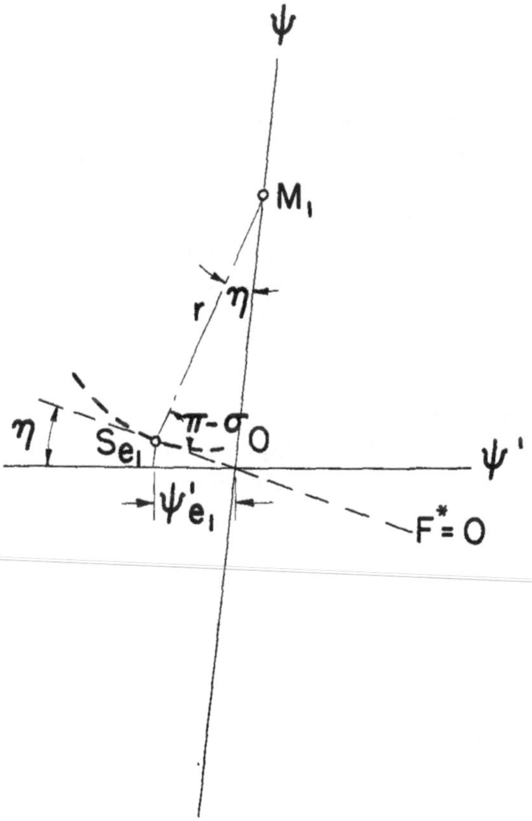

FIG. 21

The distance Se_10 is then purely a function of κ and has its maximum value when $\kappa = 1/D$; therefore,

$$(Se_10)_{\max} = \frac{1}{\sqrt{1 - D^2}} = \frac{1}{\sin\sigma} \tag{69}$$

For $\kappa = 1/D$ the angle $\eta = 90°$. Equation (66a) then becomes

$$Se_10 = (Se_10)_{\max} \cdot \cos\left(\frac{\pi}{2} - \eta\right) \tag{70}$$

and the locus of the point Se_1 is a circle passing through the points $(\psi_s, \psi'_s) = (0, 0)$ and $(\psi_s, \psi'_s) = (1, 0)$ with a diameter equal to

$$\frac{1}{\sin \sigma} = \frac{1}{\sqrt{1 - D^2}}$$

as shown in Fig. 22. The locus of the point Se_2 may be obtained by similar considerations.

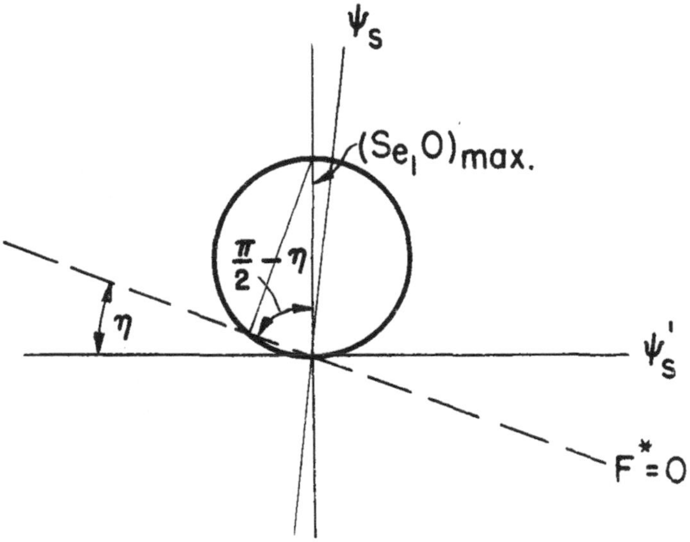

FIG. 22

Figs. 23a, b show the regions of end points and starting points in systems A and B for switch lines corresponding to various values of κ (the subscript s at the axes signifies that we are concerned *only* with a locus of switch points as illustrated in 23c).

The locus of Sr_1, which together with Se_1 characterizes the region of rest points, may be determined. Let the time required for motion from Sr_1 to Se_1 be $\nu\tau_1$. Referring to Fig. 24, the angle in the phase plane corresponding to this time is

$$\sphericalangle\, Se_1\, M_1\, Sr_1 = \nu\tau_1 \tag{71}$$

and

$$r_2 = r_1 e^{+\frac{D}{\nu}\nu\tau_1} \tag{72}$$

From Fig. 24 we see that

$$r_1 = \frac{\sin(\sigma - \eta)}{\sin(\pi - \sigma)} = \frac{\sin(\sigma - \eta)}{\sin \sigma} \tag{73}$$

and that

$$\frac{r_2}{r_1} = \frac{\sin(\pi - \sigma)}{\sin\{\pi - (\pi - \sigma + \eta) - (2\pi - \nu\tau_1 - \eta)\}} = \frac{\sin \sigma}{\sin(\nu\tau_1 + \sigma)} \tag{74}$$

The combination of equations (72) and (74) yields

$$\sin(\nu\tau_1 + \sigma) = e^{-\frac{D}{\nu}(\nu\tau_1)} \sin\sigma \tag{75}$$

Equation (75) is a transcendental relationship for the time $\nu\tau_1$, solely dependent upon D.

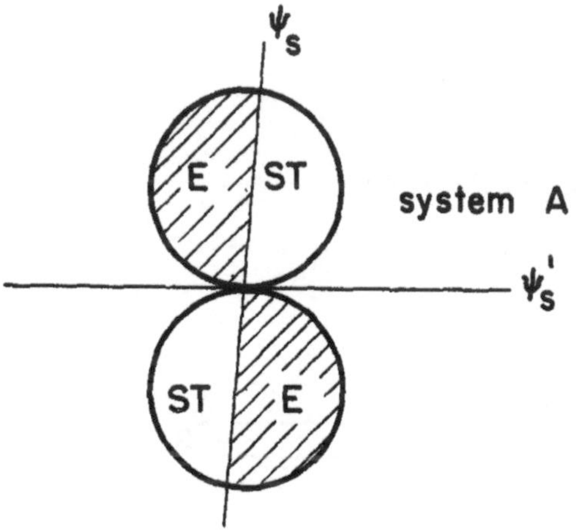

FIG. 23a

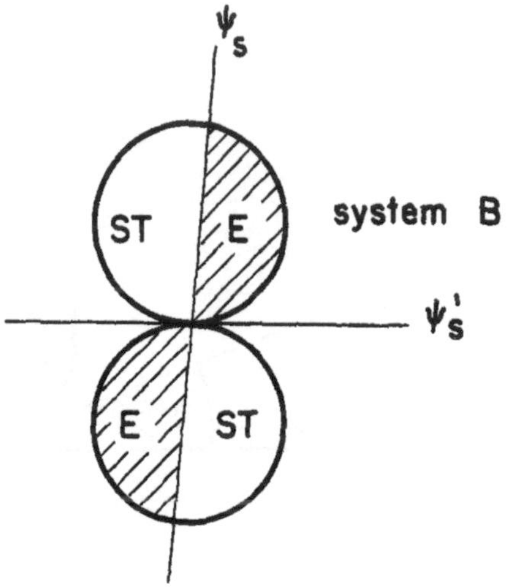

FIG. 23b

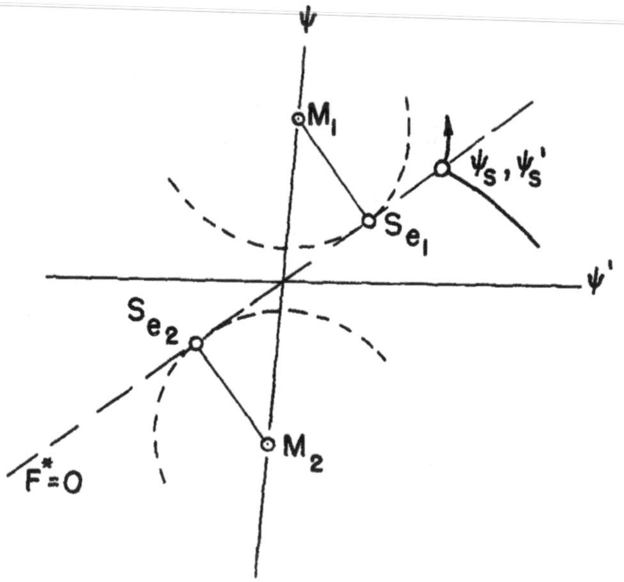

FIG. 23c

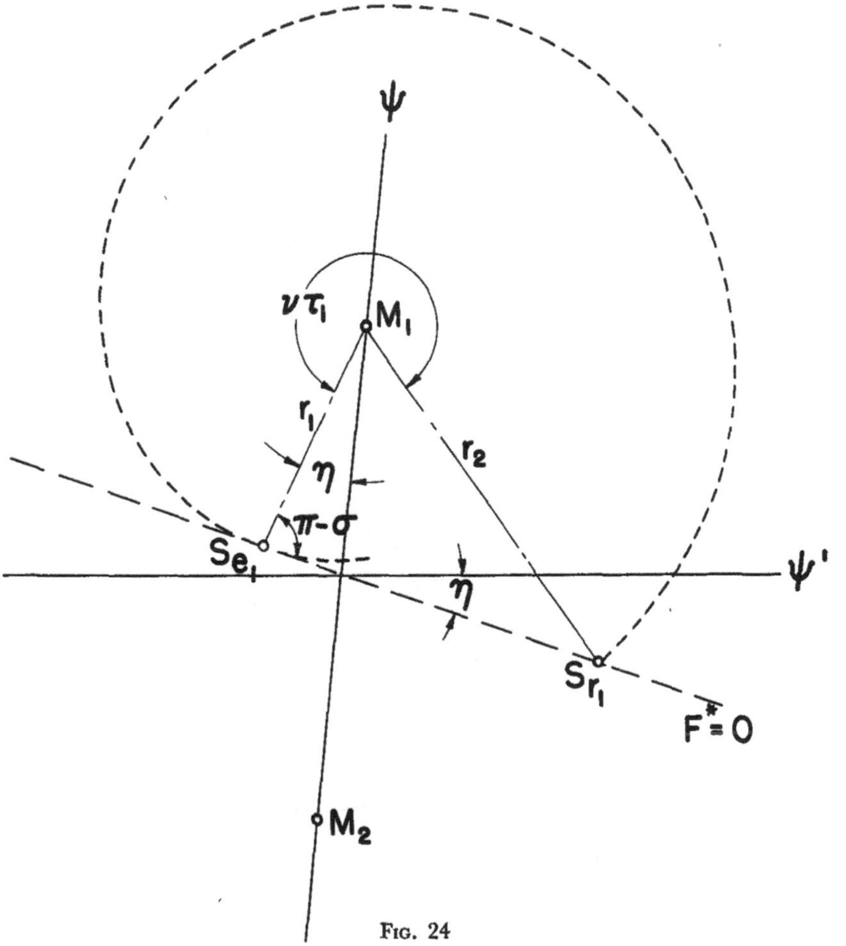

FIG. 24

It may be noted that $\nu\tau_1$ gives the maximum value of the time interval between switch points in system B, a fact evident from observation of Fig. 24. Table II below gives values of $\nu\tau_1$ for various values of D.

Table II

D	$\nu\tau_1$	$\tau_1{}^\circ$
0.05	5.50	316
0.10	5.23	302
0.20	4.85	290

$\nu = \sqrt{1 - D^2}$

Fig. 24 shows Sr_10 to be

$$Sr_10 = \frac{\sin(2\pi - \nu\tau_1 - \eta)}{\sin(\nu\tau_1 + \sigma)} = \frac{-\sin(\nu\tau_1 + \eta)}{\sin(\nu\tau_1 + \sigma)} \tag{76}$$

For fixed D the value of Sr_10 depends on κ. The maximum value of Sr_10 is found by differentiation.

$$\frac{d(Sr_10)}{d\kappa} = \frac{-1}{\sin(\nu\tau_1 + \sigma)} \frac{d}{d\kappa} [\sin(\nu\tau_1 + \eta)] = 0$$

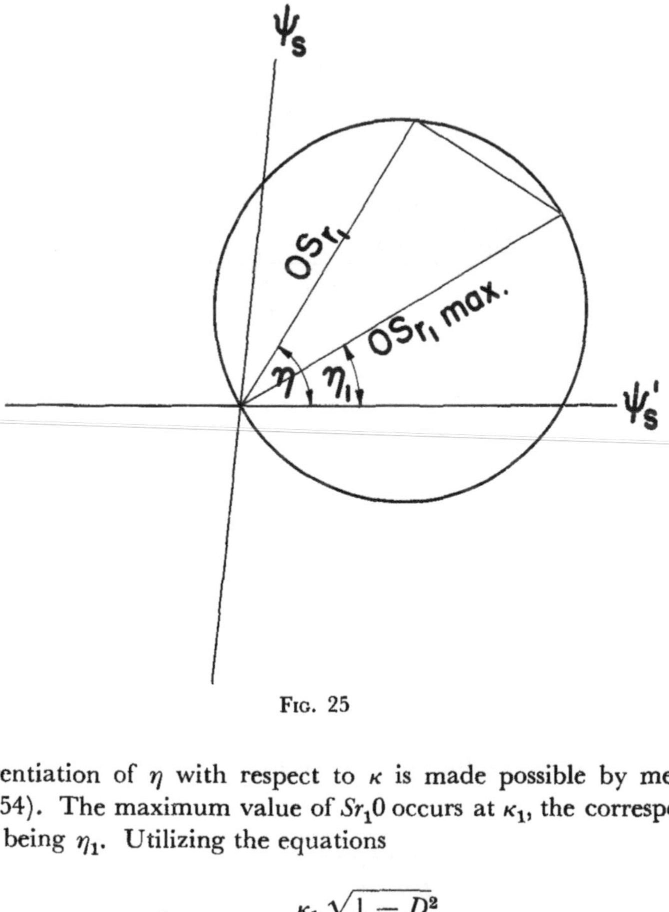

FIG. 25

The differentiation of η with respect to κ is made possible by means of equation (54). The maximum value of Sr_10 occurs at κ_1, the corresponding value of η being η_1. Utilizing the equations

$$\sin \eta_1 = \frac{\kappa_1 \sqrt{1 - D^2}}{\sqrt{1 - 2\kappa_1 D + \kappa_1^2}}$$

$$\cos \eta_1 = \frac{1 - \kappa_1 D}{\sqrt{1 - 2\kappa_1 D + \kappa_1^2}}$$

it will be found that

$$\kappa_1 = -\frac{\cos \nu\tau_1}{\cos(\nu\tau_1 + \sigma)} \tag{77}$$

In addition we find that for this value of κ_1 the root

$$\sqrt{1 - 2\kappa_1 D + \kappa_1^2} = \frac{\sin \sigma}{\cos(\nu\tau_1 + \sigma)} \tag{78}$$

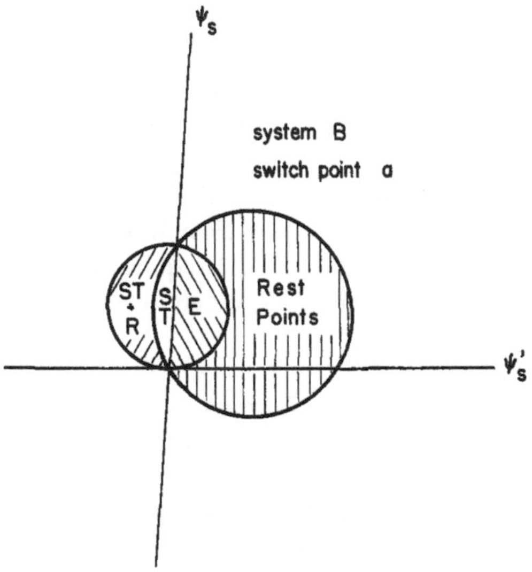

FIG. 26a

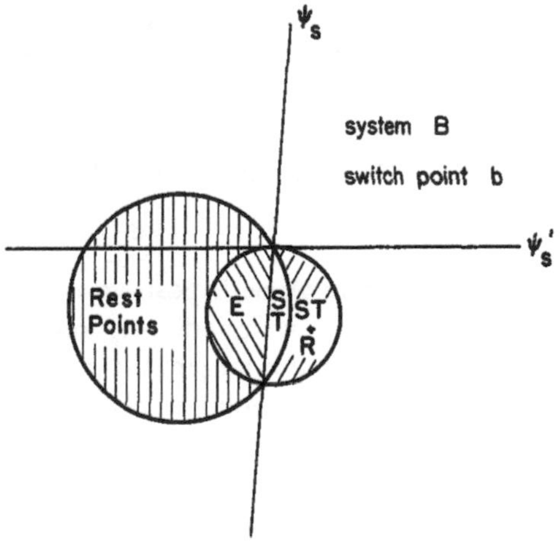

FIG. 26b

and therefore

$$\left.\begin{aligned} \sin \eta_1 &= -\cos \nu\tau_1 \\ \cos \eta_1 &= -\sin \nu\tau_1 \end{aligned}\right\} \tag{79}$$

Finally by introducing (79) in (76), the maximum value of Sr_10 is

$$(Sr_10)_{\max} = \frac{+1}{\sin(\nu\tau_1 + \sigma)} \tag{80}$$

and

$$\begin{aligned} \frac{Sr_10}{(Sr_10)_{\max}} &= -\sin(\nu\tau_1 + \eta) \\ &= -(\sin \nu\tau_1 \cos \eta + \cos \nu\tau_1 \sin \eta) \\ &= \cos(\eta_1 - \eta) \end{aligned} \tag{81}$$

Equations (80) and (81) show that the locus of Sr_10 is a circle with radius $\frac{1}{\sin(\nu\tau_1+\sigma)}$. Since this circle must include the points $(\psi_s, \psi'_s = 0, 0)$ and $(\psi_s, \psi'_s = 1, 0)$ for $\kappa = 0$ and $\kappa = \infty$, the construction may be easily performed (see Fig. 25).

Fig. 26a shows the region of rest, starting, and end points for system B with switch points of type *a*. The analogous case for switch points of type *b* is shown in Fig. 26b. The regions indicated by $ST + R$ in these figures characterize starting points which are rest points as well.

4.5. Periodic motions. At the end of Sec. 4.2 of the present chapter the statement was made that periodic motions exist, and an example was presented. We are now in a position to study in more detail the conditions for which periodic motions occur.

Fig. 27a depicts a phase curve for system A with negative κ. If the motion starting at S_1 originates at a distance d_1 from Se_2, then it is clear from the figure that for the point S_2 $d_2 < d_1$. A periodic motion would require that $OS_1 = OS_2$; in other words there is one switch point S_1 for which

$$d_1 - OSe_2 = d_2 + OSe_2$$

For a system A with positive κ as illustrated in Fig. 27b there is no such possibility for a periodic motion to occur since Se_2 lies on the same side of the origin as the larger distance d_1. Bilharz[1] has given the mathematical proof for the foregoing analysis. He studied only system A, but devoted all his work to periodic solutions and to the convergence of motion toward these periodic solutions.

The switch point coordinates of the periodic motion must now be determined. Fig. 17 shows that the period consists of two half periods with equal length. If this period is $2\tau_p$, then for a periodic motion

$$\begin{aligned}\psi_p(0) &= -\psi_p(\tau_p)\\ \psi'_p(0) &= -\psi'_p(\tau_p)\end{aligned} \tag{82}$$

Using equations (50), (51), and (56), we find for the periodic solution, after some simple but rather lengthy transformations, that

$$\psi_p = \mp \operatorname{sgn} F^* \frac{\dfrac{D}{\sqrt{1-D^2}}\sin\nu\tau_p - \sinh D\tau_p}{\cos\nu\tau_p + \cosh D\tau_p} \tag{83a}$$

and

$$\psi'_p = \pm \operatorname{sgn} F^* \frac{1}{\sqrt{1-D^2}} \cdot \frac{\sin\nu\tau_p}{\cos\nu\tau_p + \cosh D\tau_p} \tag{83b}$$

[1] H. Bilharz, Über eine gesteuerte eindimensionale Bewegung, *Zeitschrift für Angewandte Mathematik und Mechanik*, Vol. 22 (1942) 206–215.

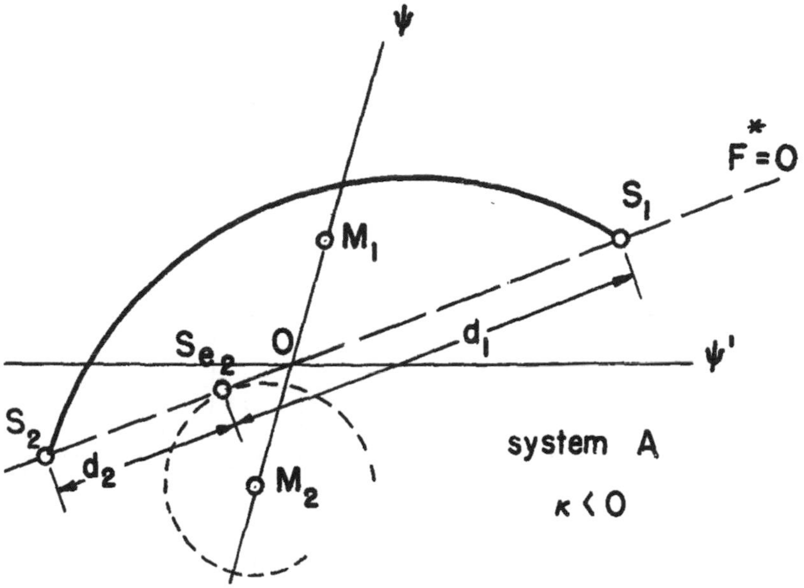

FIG. 27a

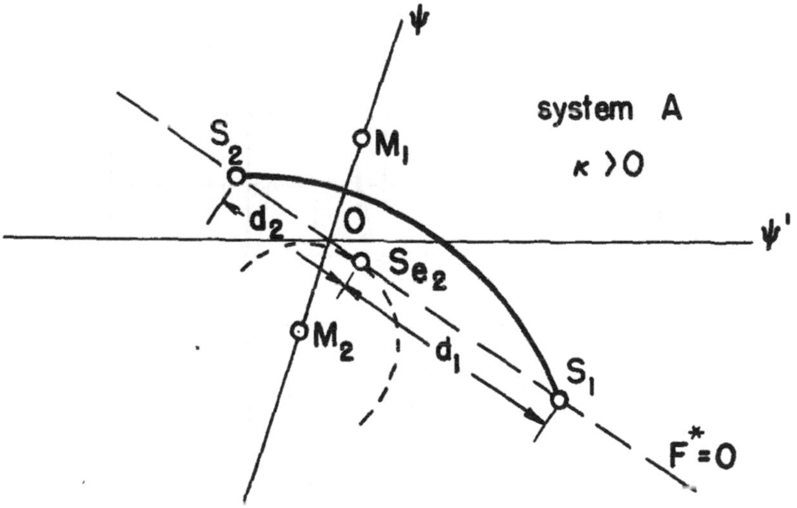

FIG. 27b

where $\cos \sigma = -D$ and $\sin \sigma = \nu = \sqrt{1 - D^2}$. Because ψ_p and ψ'_p are switch point coordinates, we may write

$$\psi_p + \kappa \psi'_p = 0 \tag{84}$$

Therefore the relationship between the length of the period $(2\tau_p)$ and the coefficient κ of the system is known.

Now it should be clear that periodic motions may exist for system A, with $\kappa < 0$ only and for system B with $\kappa > 0$ only. The reader may verify

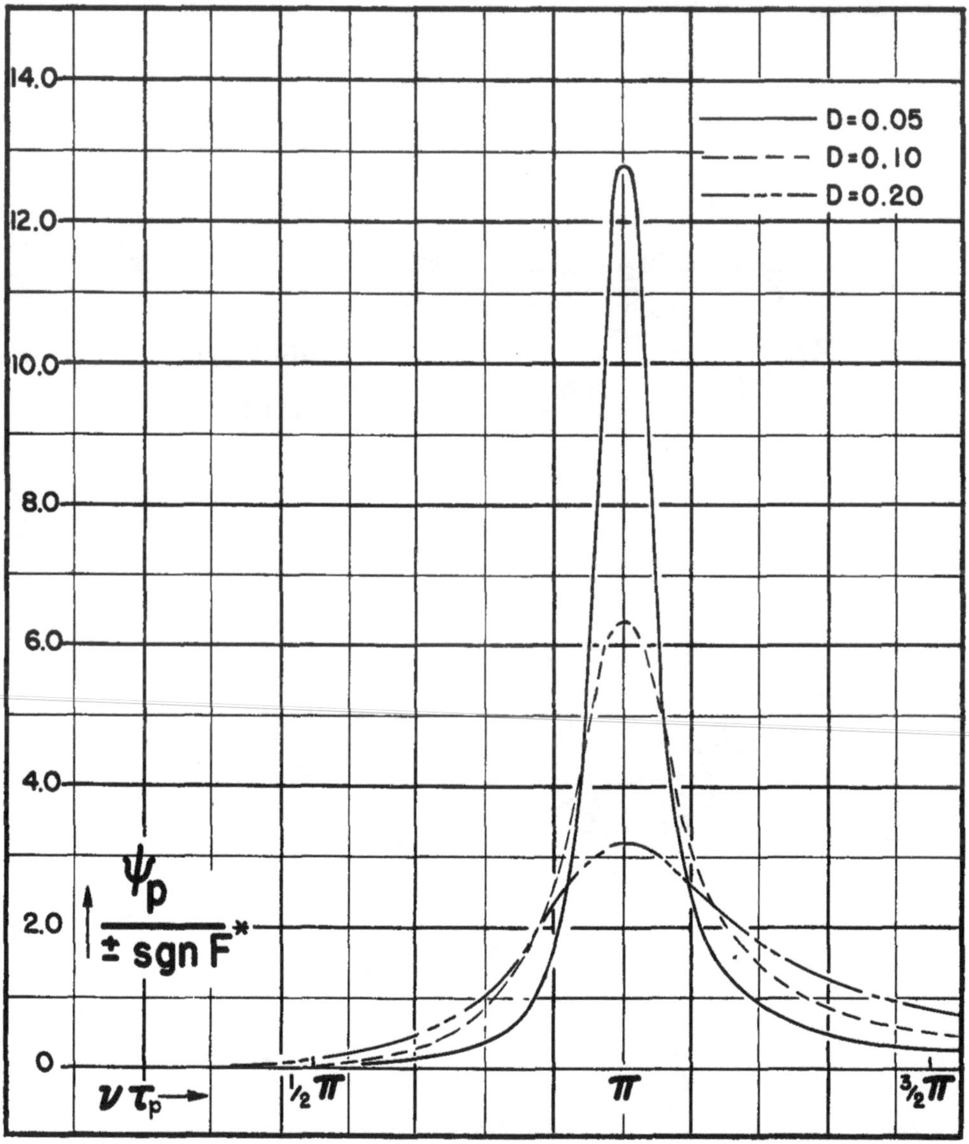

FIG. 28a

this for system B by sketching figures analogous to those shown in Figs. 27a, b for system A.

Curves for switch point coordinates ψ_p and ψ_p', and the coefficient κ are represented in Figs. 28a, b, c, versus $\nu\tau_p$ for three different values of the parameter D. The curves will be used primarily in an inverse manner; that is to say that for a given D and κ the length of the period and the

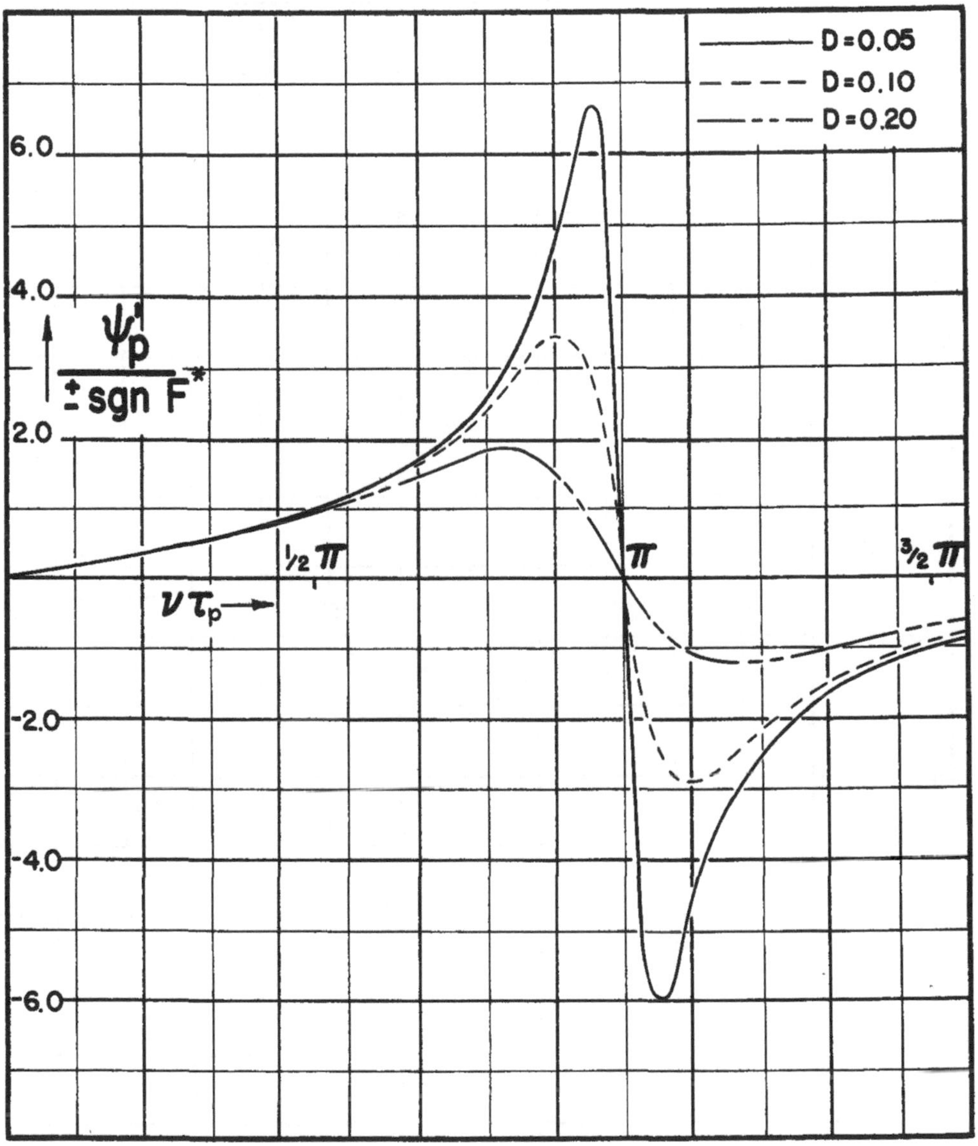

FIG. 28b

coordinates of the switch points will be found. For every value of D there is a region with $\nu\tau_p > \pi$ for which κ is not a single valued function. It can be shown that *only* the solution with the lower value of $\nu\tau_p$ is a stable one. Fig. 29 shows the coordinates ψ_p and ψ_p' in a different representation. This figure shows clearly the influence of D upon the periodic motion.

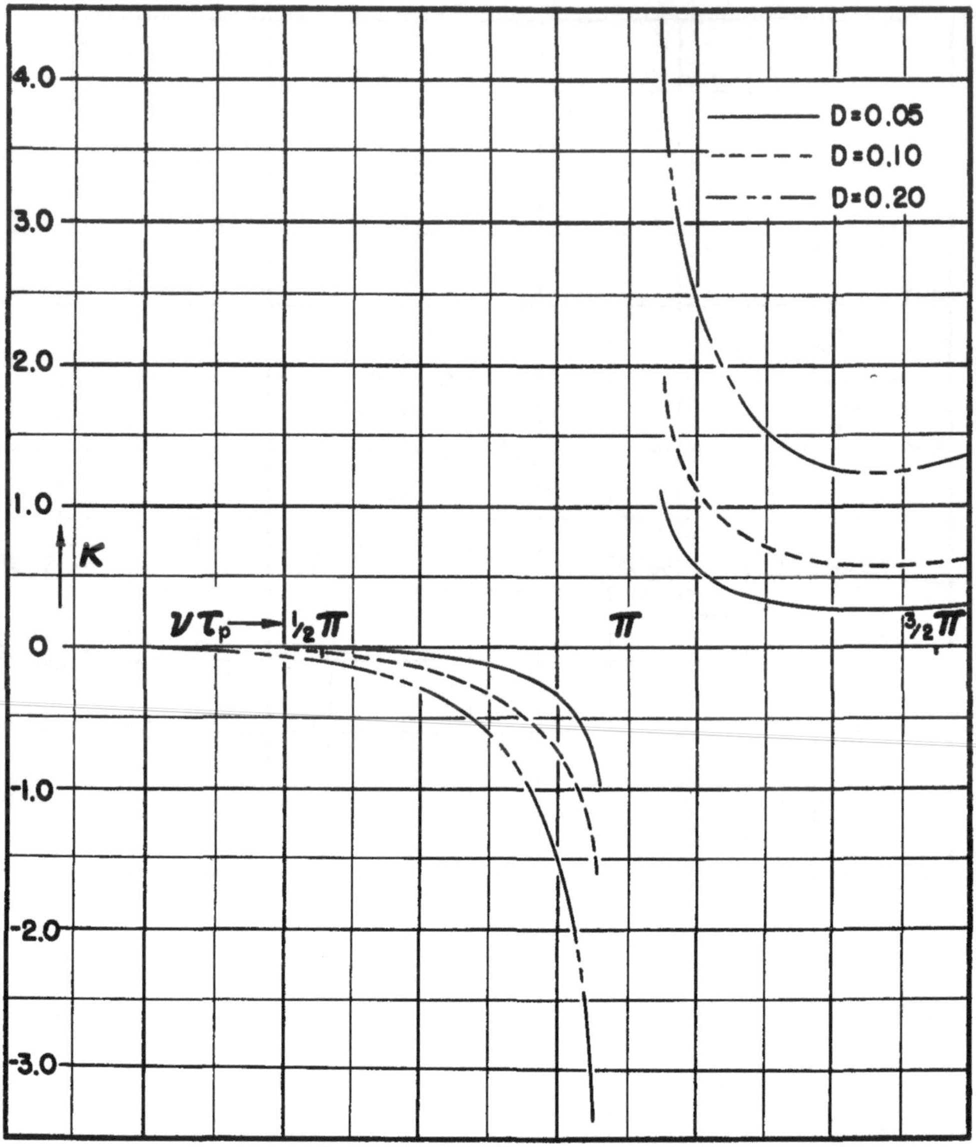

FIG. 28c

In system A, owing to the location of the points M_1 and M_2 with respect to the half plane in which the spirals appear, every interval is smaller than π/ν (readily seen by observing Fig. 13); thus the period $(2\tau_p)$ of the motion is smaller than $2\pi/\nu$. In ordinary time units (the notation $\tau = \omega t$ was

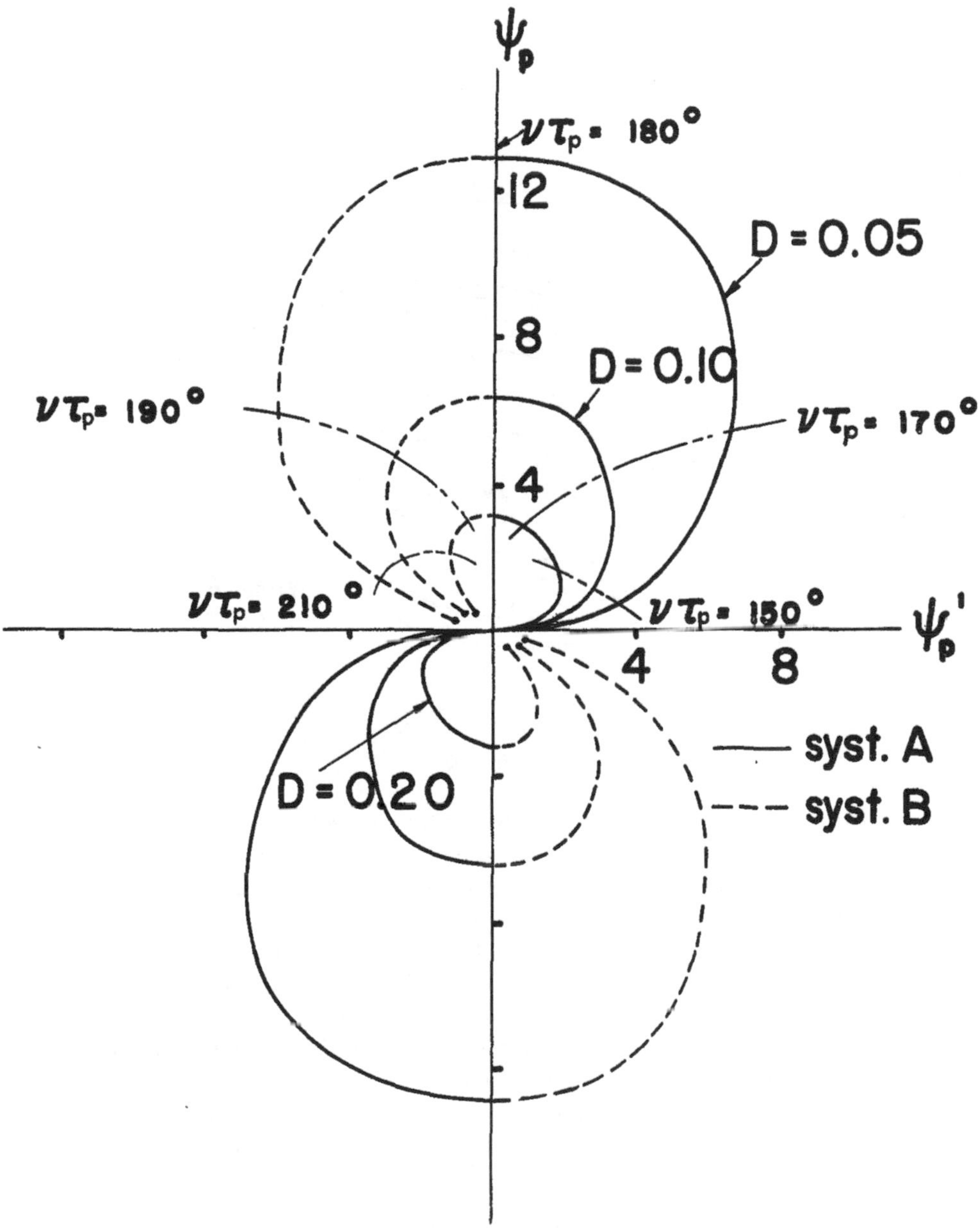
ψp
ντp = 180°
12
D = 0.05
8
D = 0.10
ντp = 190°
ντp = 170°
4
ντp = 210°
ντp = 150°
4
8
ψp'
syst. A
syst. B
D = 0.20

FIG. 29

introduced on p. 18) this fact means that the period is smaller than $2\pi/\omega\nu$; the frequency, higher than $\omega\nu/2\pi$. In system B the frequency is lower than $\omega\nu/2\pi$.

4.6. Complete survey of the possible motions.

System A; positive κ. In an ideal control mechanism all phase curves will lead to an end point (refer again to Figs. 13 and 15). In an imperfect

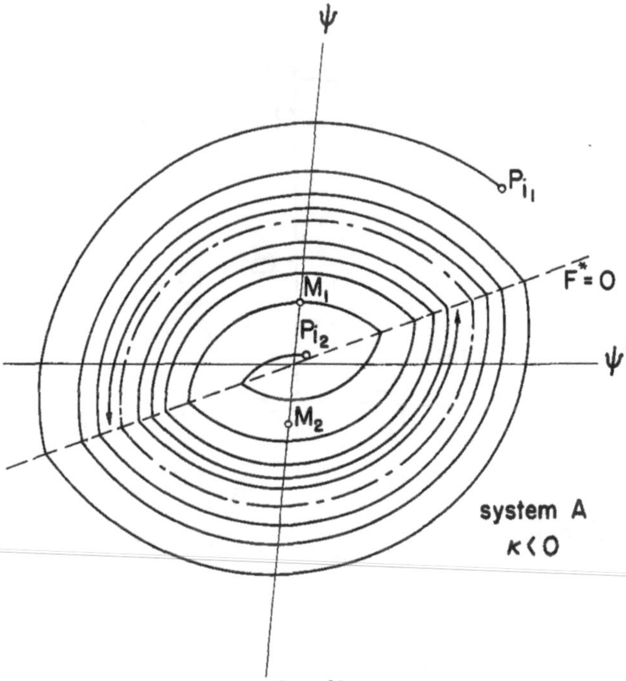

FIG. 30

control mechanism the motion passes to one of very high frequency (Fig. 16). The regions for the end points are given in Fig. 23a.

System A; negative κ. In this system periodic motions exist. In Fig. 30 two phase curves, one starting at the point Pi_1 and the other at the point Pi_2 converge upon a limit cycle.[1] The former curve approaches the periodic motion (shown by a dashed line) with *decreasing* amplitude; the latter curve approaches the periodic motion with *increasing* amplitude. In system A all intervals are less than half the period of the uncontrolled motion. This result corresponds with the statement made earlier (see page 11) that in system A the control force (or moment) reinforces the restoring couple.

System B; positive κ. Again periodic motions are possible, one such motion being shown by the dashed line in Fig. 31a. As expected, phase

[1] The periodic motion is a limit cycle as defined by Poincaré (see N. Minorsky, *Introduction to Non-linear Mechanics*, Edwards Ann Arbor, 1947, p. 62ff.)

curves originating in the region exterior to the periodic curve will converge to the periodic curve with decreasing amplitude, phase curves originating in the region within the periodic curve will tend towards it with increasing amplitude. In this latter case one important exception should be noted: phase curves starting within the shaded region near the origin result in motions with control at rest. Figs. 26a, b describe all possible motions near

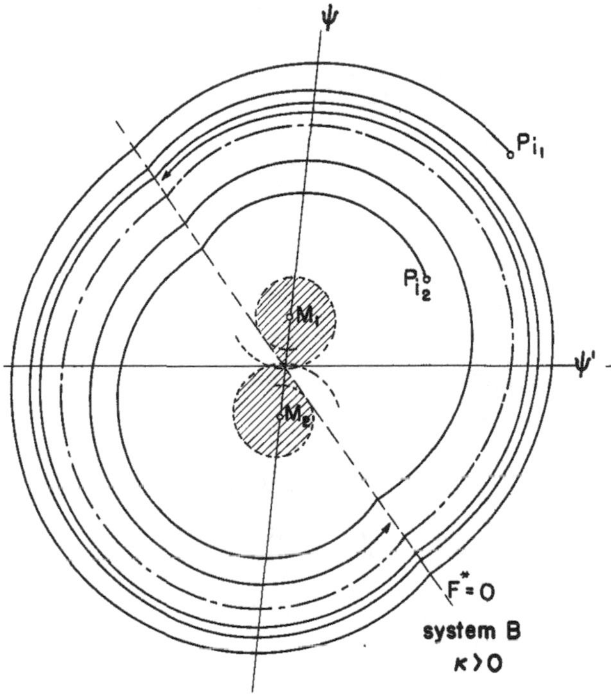

FIG. 31a

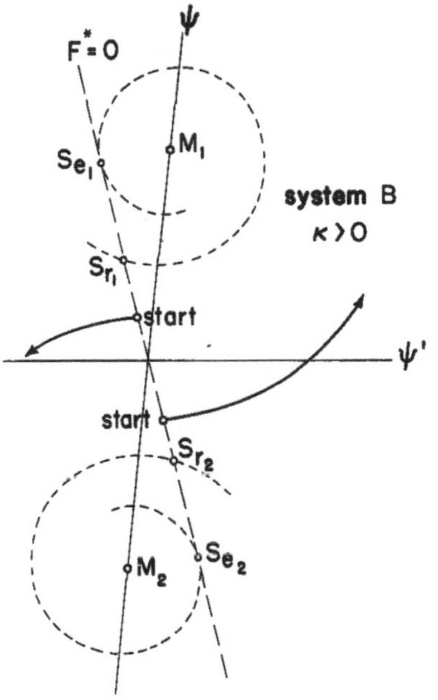

FIG. 31b

the origin for system B with positive κ, classifying them by their switch points. Fig. 31b shows clearly that for large positive values of κ there exists, as indicated in Fig. 26a, a region near the origin in which starting points occur and the control operates.

System B; negative κ. Owing to considerations similar to those employed in the discussion of Figs. 27a, b, no periodic motions exist. However, motions may occur which lead to rest points or to end points. In Fig. 32 two phase curves starting at points Pi_1 and Pi_2 are drawn as examples of those motions which would degenerate into motions with control at rest. The shaded region in the phase plane includes all phase curves which eventually lead to switch points lying between Se_1 and Se_2. We have seen that all such switch points are end points, at which all motions cease in the ideal case at least. However, in the practical case, time lags almost always exist. Therefore a high-frequency motion will begin in what would have normally been termed an end point. An example of this type is given in

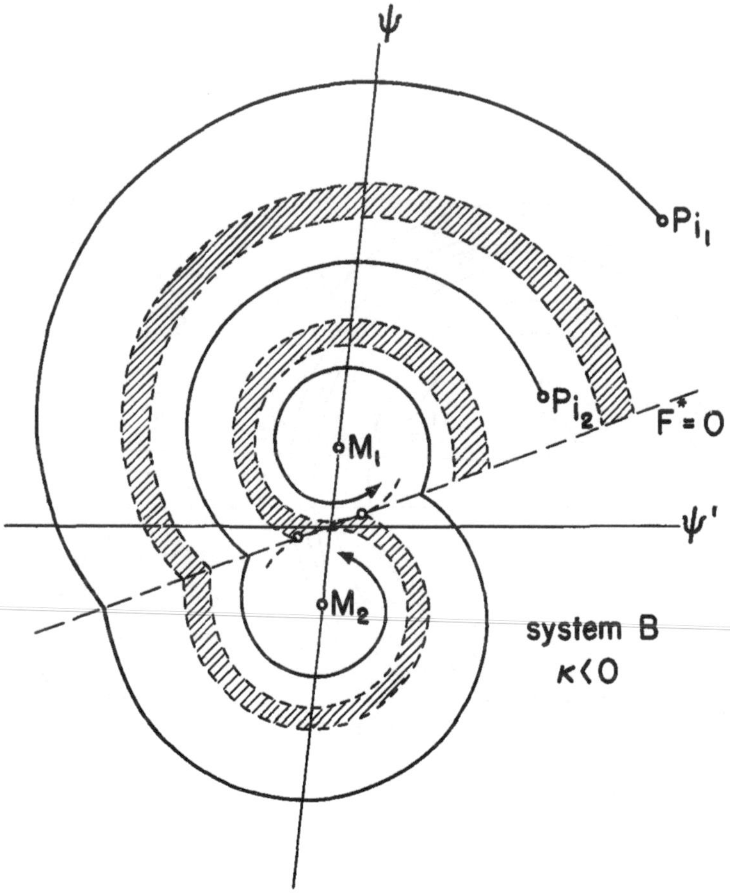

FIG. 32

Fig. 33. The high-frequency motion will not tend towards the origin, as for system A, but it will grow first in amplitude and then become a motion with control at rest.

4.7. Selection of the system and the coefficient κ for a given mechanical problem. In the preceding sections the qualities of the two different control systems A and B, have been examined and explained. The final conclusion is that only system A with positive values of κ can improve the damping qualities of the motion. In this system every motion will tend towards an end point, if an ideal control mechanism could be realized.

However, as previously indicated time lags always exist. The motion will not be undefined at the end point, but will pass to a very-high-frequency small-amplitude motion (see Fig. 16) with an average path leading towards

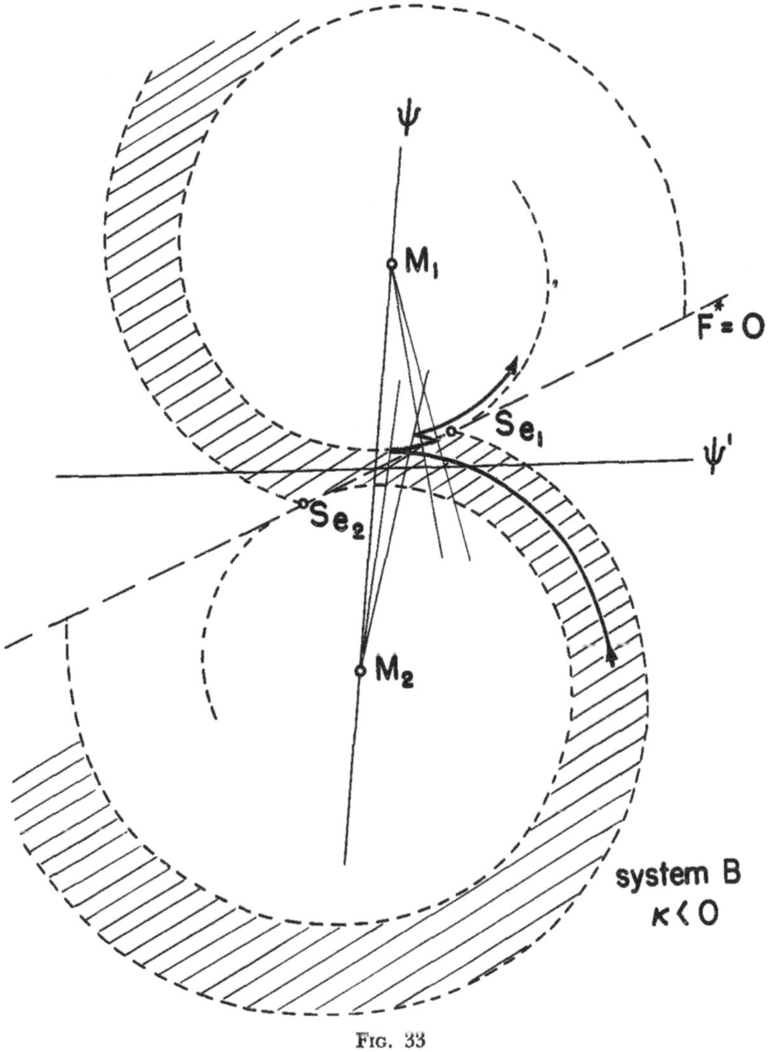

FIG. 33

the origin ($\psi = 0, \psi' = 0$). Subsequent investigations (Chap. 6) will prove that the final motion will be a high-frequency small-amplitude periodic one about the origin. The magnitude of this amplitude depends upon the character and degree of the time lag.

Now that the system and the sign of the coefficient κ have been chosen, only the absolute value of κ has to be determined. First it is necessary to remind the reader that in Sec. 3.1. new coordinates were introduced in order to arrive at the essential parameters of the problem. Therefore, in any practical problem equation (16) must be considered. The new coefficients, defined by equation (18), are advantageous then only from the viewpoint of the computer. Equations (22) and (23) determine the units; thus the ratio of the magnitudes of the anticipated disturbances to the size of the forces of the control mechanism is also fixed. The value $\psi = 1$ corresponds to $\phi = N\beta_0/\omega^2$. We would prefer that the phase curves lead to an end point quickly in order to initiate a rapid average motion towards $\psi = 0$, $\psi' = 0$. Therefore, $N\beta_0$ must be chosen so that the initial values (ϕ_i, ϕ_i') have a point (ψ_i, ψ_i') near the region of the end points (that means $|\psi_i| < 1, |\psi_i'| < 1$ as is shown in Fig. 23a). From equation (63) we see that for a rapid trend towards zero, κ should be small, But if κ is chosen very small, then the region of end points (measured in the ψ, ψ' system)

$$Se_1 0 = \frac{\kappa}{\sqrt{1 - 2\kappa D + \kappa^2}} \tag{85}$$

is also very small. Thus, in order to maintain κ at small values it will be advisable to make $N\beta_0/\omega^2$ so large that the points giving the initial disturbance lie about a distance $Se_1 0$ from the origin.

5. COMPLETE THEORY OF THE MOTION WITH VELOCITY CONTROL[1]

5.1. The representation of the motion in the phase space. This type of motion has been shown to be governed by equation (42):

$$\bar{\psi}'' + 2D\bar{\psi}' + \bar{\psi} = \bar{\beta} \tag{86a}$$

$$\bar{\beta}' = \mp \operatorname{sgn} F_1^* \tag{86b}$$

with the control function F_1^* given by

$$F_1^* = \bar{\psi} + \kappa\bar{\psi}' + \mu\bar{\beta} \tag{87}$$

Corresponding to equations (12a, b), we may define two control systems:

$$\left.\begin{aligned} \bar{\beta}' &= - \operatorname{sgn} F_1^* \quad . \quad . \quad . \quad . \quad \text{control system } C_1 \\ \bar{\beta}' &= + \operatorname{sgn} F_1^* \quad . \quad . \quad . \quad . \quad \text{control system } C_2 \end{aligned}\right\} \tag{88}$$

The motion is completely determined by the three initial values $\bar{\psi}_i$, $\bar{\psi}_i'$, and $\bar{\beta}_i$. In the mth interval the solution will be given by $\bar{\psi}_m(\tau_m)$ and $\bar{\beta}_m(\tau_m)$:

$$\left.\begin{aligned} \bar{\psi}_m(\tau_m) &= 2C_m e^{-D\tau_m} \cos(\nu\tau_m + \varepsilon_m) \pm 2D \operatorname{sgn} F_1^* + \bar{\beta}_{mi} \\ &\quad \mp (\operatorname{sgn} F_1^*)\tau_m \\ \bar{\beta}_m(\tau_m) &= \bar{\beta}_{mi} \mp (\operatorname{sgn} F_1^*)\tau_m \end{aligned}\right\} \tag{89}$$

The first derivative of $\bar{\psi}_m(\tau_m)$ is given by

$$\bar{\psi}_m'(\tau_m) = 2C_m e^{-D\tau_m} \cos(\nu\tau_m + \varepsilon_m + \sigma) \mp \operatorname{sgn} F_1^* \tag{90}$$

where $\cos\sigma = -D$, $\sin\sigma = \sqrt{1 - D^2} = \nu$

Using equations (89) and (90) we may construct the path of the motion for successive intervals (see p. 20), once the initial values for $\bar{\psi}_i$, $\bar{\psi}_i'$, and $\bar{\beta}_i$ have been given. As was pointed out earlier, however, such a procedure is lengthy and tedious. Again it appears worthwhile to seek a different representation of the motion.

[1] This problem was treated first by K. Klotter and H. Hodapp in: Über Bewegungen eines Schwingers unter dem Einfluss von Schwarz-Weiss-Steuerungen. III. Bewegungen eines Schwingers mit Laufgeschwindigkeitszuordnung (ohne und mit starrer Rückführung), *Zentrale für wissenschaftliches Berichtswessen der Luftfahrtforschung des Generalluftzeugmeisters* (*ZWB*), Untersuchungen und Mitteilungen Nr. 1328, Berlin, November 1, 1944. In this chapter an entirely new representation is given, one which avoids the auxiliary diagrams with isochronal curves of the German report.

As in the theory of the position control, it is convenient to avoid the use of time as the independent variable describing the motion. Instead the time will be eliminated by introducing the derivative $\bar{\psi}'$ as a third variable in addition to $\bar{\psi}$ and $\bar{\beta}$. In "phase space" with coordinate axes $\bar{\psi}$, $\bar{\psi}'$, and $\bar{\beta}$, the initial values $\bar{\psi}_i$, $\bar{\psi}_i'$, and $\bar{\beta}_i$ determine a single point. The motion is described by a continuous trajectory.

In this representation time plays the role of a parameter. Because $\bar{\beta}$ depends linearly on the time (see equation (89)), the $\bar{\beta}$ values at two points of the motion allow ready determination of the time required to go from one point to the other.

The points in the phase space for which the control function F_1^* equals zero are located on a predetermined plane having fixed κ and μ. It should be clear that such a representation of the motion possesses great advantages over other possible methods. Thus we must determine how to construct the phase curve in the phase space.

Let us define two function χ and χ' so that

$$\left.\begin{aligned} \chi_m(\tau_m) &= \bar{\psi}_m(\tau_m) - \bar{\beta}_m(\tau_m) = 2C_m e^{-D\tau_m} \cos(\nu\tau_m + \varepsilon_m) \pm 2D \operatorname{sgn} F_1^* \\ \chi_m'(\tau_m) &= \bar{\psi}_m'(\tau_m) - \bar{\beta}_m'(\tau_m) = 2C_m e^{-D\tau_m} \cos(\nu\tau_m + \varepsilon_m + \sigma) \end{aligned}\right\} \tag{91}$$

Comparison of the analytical expressions for $\chi(\tau_m)$ and $\chi'(\tau_m)$ with those for ψ and ψ' in the theory of position control (see equations (50), (51), (47), (52)) shows that they have the same analytical form. This means that in an oblique coordinate system (χ, χ') as shown in Fig. 34, the paths of the points $\chi(\tau)$, $\chi'(\tau)$ are logarithmic spirals around centers given by

$$\chi = \pm 2D \operatorname{sgn} F_1^* \tag{92}$$

with a radius

$$r_m = 2C_m e^{-D\tau_m} \sin \sigma \tag{93}$$

For example, suppose that the system C_1 has been chosen with an interval for which $F_1^* > 0$. The point (χ, χ') traces a spiral around $M_1 = +2D$ in the χ, χ' plane (see Fig. 34). For this reason the construction of phase curves in a χ, χ', $\bar{\beta}$ space is much less difficult than in a $\bar{\psi}$, $\bar{\psi}'$, $\bar{\beta}$ space. Therefore we will confine ourselves to a χ, χ', $\bar{\beta}$ space, for which the phase curve in the mth interval is given by

$$r_m = 2C_m e^{-D\tau_m} \sin \sigma \tag{94a}$$

$$\bar{\beta}_m = \bar{\beta}_{mi} \mp (\operatorname{sgn} F_1^*)\tau \tag{94b}$$

where r_m is measured from the pertinent center $(\pm 2D \operatorname{sgn} F_1^*)$.

The time interval between any two points on the trajectory between two consecutive switch points may be given either by the angle between the radii from the center of the spiral to those points in the χ, χ' plane, or by the difference in the values of $\bar{\beta}$ for those points.

The transformation

$$\left.\begin{aligned} \chi &= \bar{\psi} - \bar{\beta} \\ \chi' &= \bar{\psi}' - \bar{\beta}' \\ \bar{\beta} &= \bar{\beta} \end{aligned}\right\} \tag{95}$$

has proved to be very successful for the construction of the phase curve in a

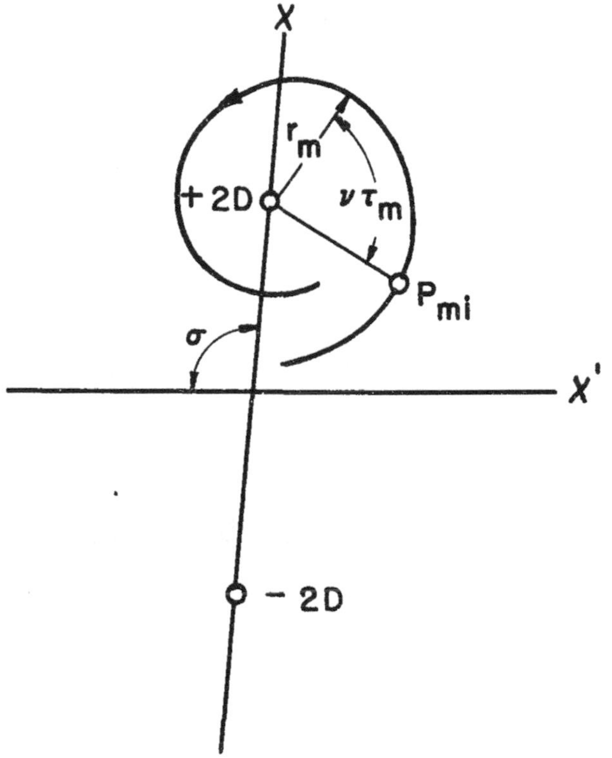

FIG. 34

single interval. It should be noted that the transformation has a special peculiarity because it involves a discontinuous function $\bar{\beta}'$.

The $\bar{\psi}$, $\bar{\psi}'$, $\bar{\beta}$ space is divided by the plane

$$F_1^* = 0 = \bar{\psi} + \kappa\bar{\psi}' + \mu\bar{\beta} \tag{96}$$

(containing the origin $\bar{\psi} = 0$, $\bar{\psi}' = 0$, $\bar{\beta} = 0$) into two "half spaces," one for $F_1^* > 0$ and the other for $F_1^* < 0$. Corresponding to each of these half spaces there is a constant value of $\bar{\beta}'$, which differs in each only in sign. Therefore equations (95) transform the two half spaces separately. As yet we have to determine if the combination of the two transformed half spaces will form one complete space, or if the result will be some distorted space with regions of gap or overlap.

In terms of the new coordinates the control function F_1^* is given by

$$F_1^* = \chi + \kappa\chi' + \bar{\beta}(1 + \mu) \mp \kappa \operatorname{sgn} F_1^* \tag{97}$$

It is obvious that $F_1^* = 0$ defines two different planes, neither of which passes through the origin ($\chi = 0$, $\chi' = 0$, $\bar{\beta} = 0$). The two half spaces, which were transformed separately, will not combine to form a complete space, but they will have regions of gap or overlap. These two possibilities depend upon the selection of the control system C_1 or C_2 as well as the sign of the coefficient κ.

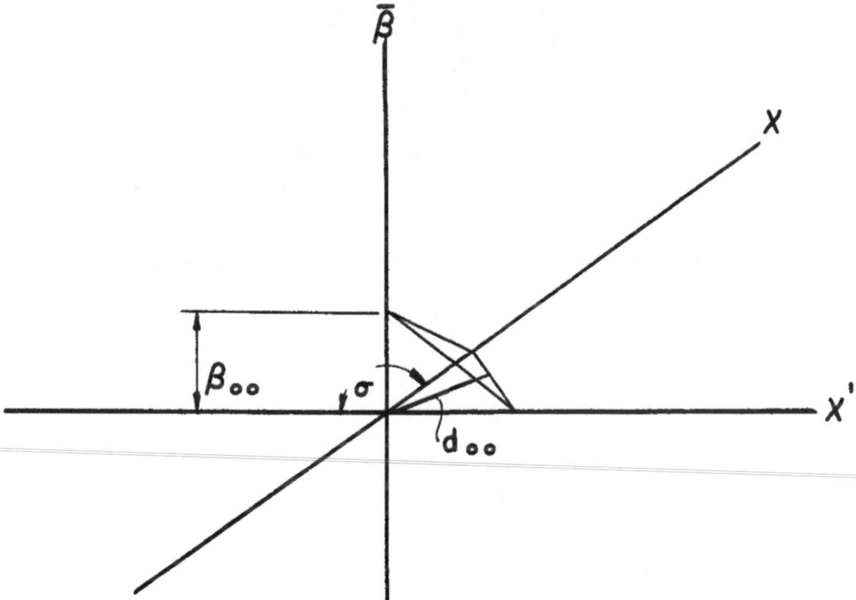

FIG. 35

The planes defined by $F_1^* = 0$ will intersect the $\bar{\beta}$ axis at the points $\bar{\beta}_{00}$

$$\bar{\beta}_{00} = \pm \frac{\kappa \operatorname{sgn} F_1^*}{1 + \mu} \tag{98}$$

and they will intersect the χ, χ' plane in the lines

$$\chi + \kappa\chi' \mp \kappa \operatorname{sgn} F_1^* = 0 \tag{99}$$

as shown in Fig. 35. (Only the upper plane $F_1^* = 0$ is shown). These lines will form an angle ζ with the negative direction of the χ' axis, where

$$\tan \zeta = \frac{\kappa\sqrt{1 - D^2}}{1 - \kappa D} \tag{100}$$

The distance of these lines to the origin d_{00} is given by

$$d_{00} = (\pm \operatorname{sgn} F_1^*) \frac{\kappa\sqrt{1 - D^2}}{\sqrt{1 - 2\kappa D + \kappa^2}} \tag{101}$$

The switching planes $F_1^* = 0$ form an angle δ with the χ, χ' plane.

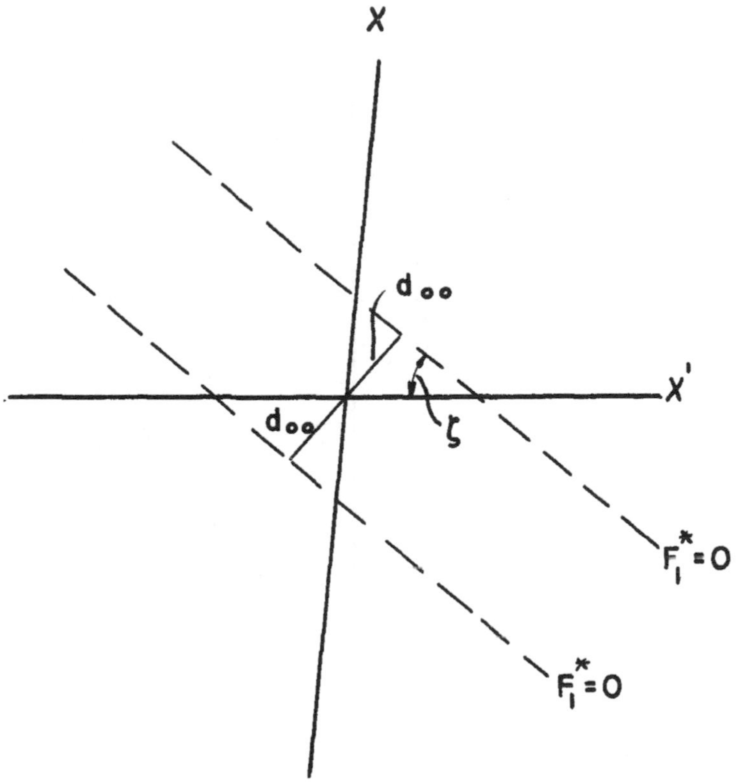

FIG. 36

$$\tan \delta = \frac{\bar{\beta}_{00}}{d_{00}} = \frac{\sqrt{1 - 2\kappa D + \kappa^2}}{(1 + \mu)\sqrt{1 - D^2}} \tag{102}$$

It is independent of the sign of F_1^*. All of the foregoing notation is shown geometrically in Figs. 35, 36, 37.

It has been found most convenient to construct the phase curve by combining successive segments obtained from equation (94) in the following manner: The phase curve will be represented by means of two projections, one in the χ, χ' plane and a second in a plane normal to the χ, χ' plane, containing the $\bar{\beta}$ axis, and normal to the switching planes. This plane intersects the χ, χ' plane in the line d_1, which contains the section d_{00} (see Figs. 37 and 38). Call this plane the d_1, β plane.

The coordinate d_1 of the phase curve in a single interval is given by

$$d_{1_m} = \chi_m \sin(\sigma - \zeta) + \chi'_m \sin \zeta \tag{103}$$

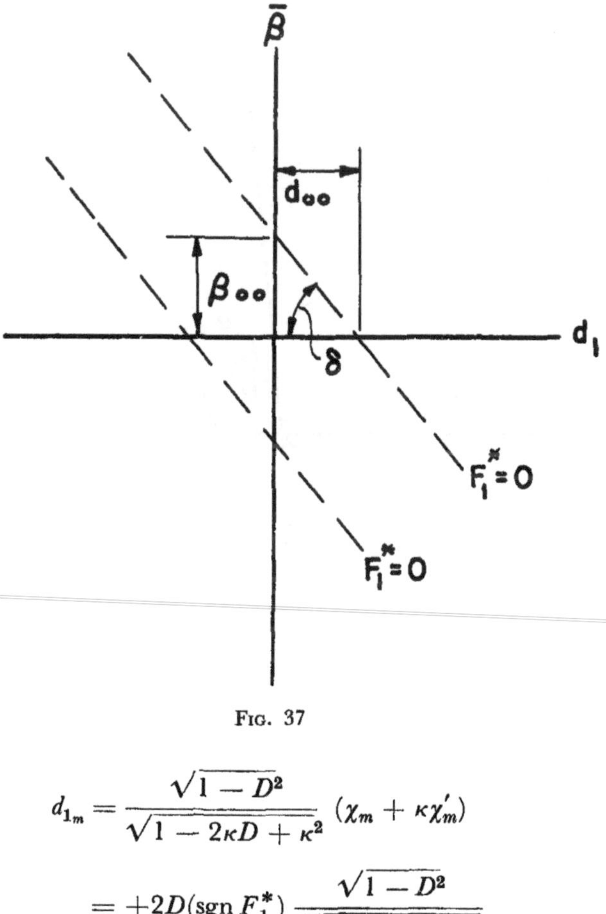

FIG. 37

or

$$d_{1_m} = \frac{\sqrt{1 - D^2}}{\sqrt{1 - 2\kappa D + \kappa^2}} (\chi_m + \kappa \chi'_m) \tag{104a}$$

$$= \pm 2D(\operatorname{sgn} F_1^*) \frac{\sqrt{1 - D^2}}{\sqrt{1 - 2\kappa D + \kappa^2}}$$

$$+ 2C_m \sqrt{1 - D^2}\, e^{-D\tau_m} \cos(\nu \tau_m + \varepsilon_m + \zeta) \tag{104b}$$

in which ζ is given by equation (100). Furthermore, since

$$\bar{\beta}_m = \bar{\beta}_{mi} \mp (\operatorname{sgn} F_1^*)\tau_m \tag{94}$$

the projection $d_{1_m}(\bar{\beta}_m)$ is a segment of a damped cosine line. Now that all relations necessary for the construction of a phase curve in a single interval have been formulated, we are in a position to investigate an example in a sequence of intervals.

Example I. Let us choose the coefficients $\kappa = 2.3$ and $\mu = 0$ (no feedback), control system C_1, the parameter $D = 0.1$, and the initial values:

$$\bar{\psi}_i = 2.1$$
$$\bar{\psi}_i' = 1.5$$
$$\bar{\beta}_i = -0.7$$

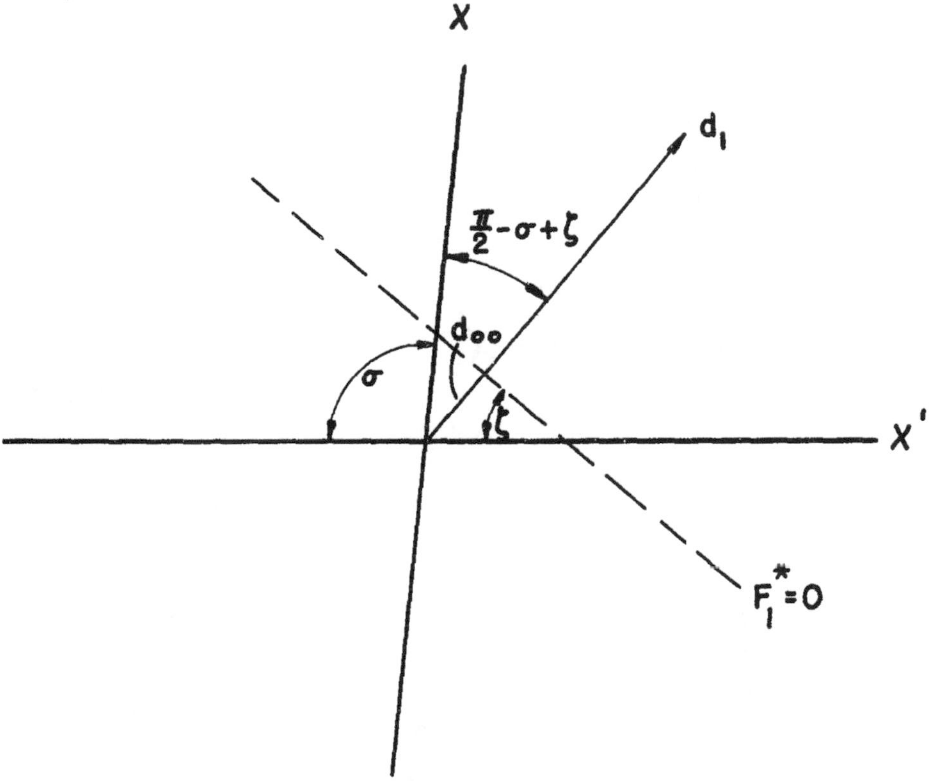

FIG. 38

Initially the control function F_1^* has the value

$$F_{1_i}^* = 2.1 + \kappa(1.5) = 5.55$$

which means that

$$\operatorname{sgn} F_{1_i}^* = +1$$

Hence

$$\chi_i = \bar{\psi}_i - \bar{\beta}_i = 2.8$$
$$\chi_i' = \bar{\psi}_i' - \bar{\beta}_i' = \bar{\psi}_i' + \operatorname{sgn} F_1^* = 1.5 + 1 = 2.5$$

The centers of the spirals in the χ, χ' plane are given by $\pm 2D \operatorname{sgn} F_1^*$. Therefore we must initiate the motion with a spiral around $M_1 = +2D$

in the χ, χ' plane. The axis d_1 forms the angle $(\pi/2 - \zeta)$ with the positive χ' axis in the χ, χ' plane. Thus according to equation (100), we find

$$\frac{\pi}{2} - \zeta = 18.5^\circ$$

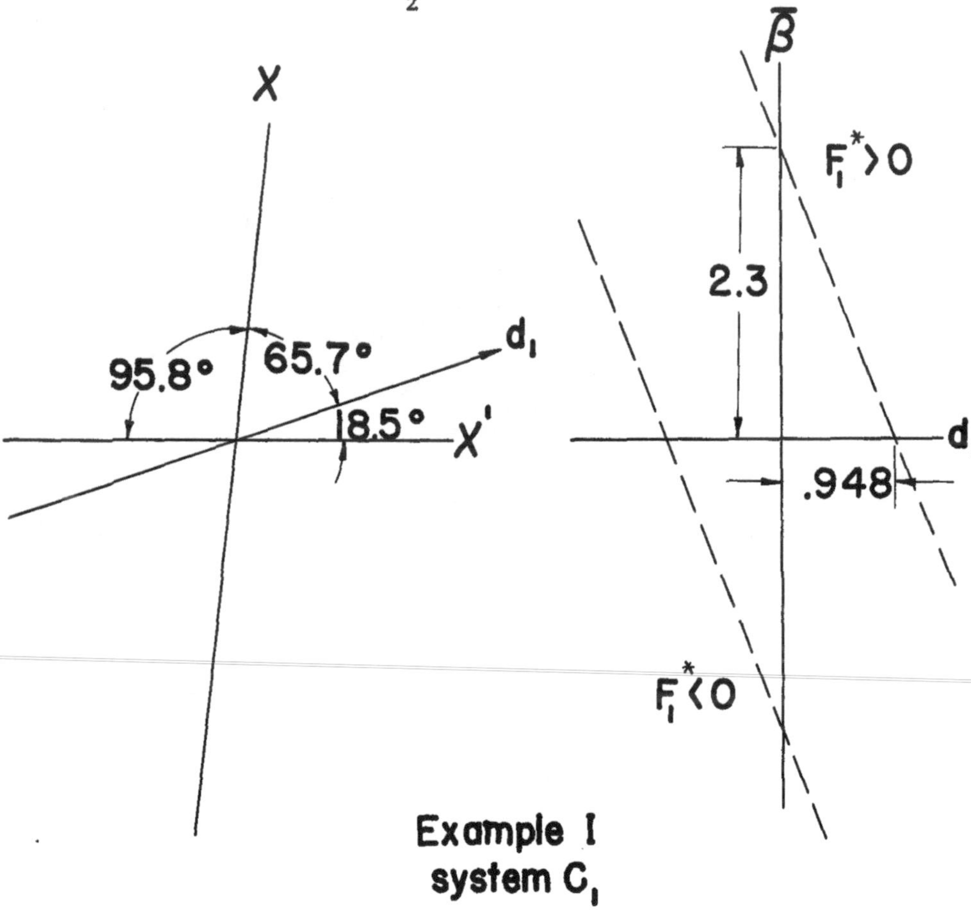

FIG. 39

The traces of the switching planes in the d_1, $\bar{\beta}$ plane are given by $\bar{\beta}_{00}$ and d_{00}.

$$\bar{\beta}_{00} = +\,2.3 \operatorname{sgn} F_1^*$$
$$d_{00} = +\,0.948 \operatorname{sgn} F_1^*$$

Now we may draw all the essential lines in the two projection planes as shown in Fig. 39, and the so-called "construction frame" is obtained. The concept of a "frame" is of extreme importance and will be more thoroughly discussed later (Sec. 5.4). For this particular example the figure shows clearly that the two half spaces do not join but are some distance apart.

The actual construction of the phase curve is begun by indicating the starting point $P_i = (\chi_i, \chi_i', \bar{\beta}_i)$ in the χ, χ' and d_1, $\bar{\beta}$ planes. In the χ, χ' plane this point is projected to the line d_1 where it is denoted as P_i. This value of d_1 is then placed in the d_1, $\bar{\beta}$ plane and together with $\bar{\beta}_i = -0.7$ indicates P_i in the d_1, $\bar{\beta}$ plane. These remarks and the remainder of the discussion regarding Example I refer to Figs. 40a, b.

Now the spiral may be sketched in the χ, χ' plane passing through P_i around the center $M_i = +2D$. In order to find the point where the phase curve will intersect the switching plane, we must draw its projection in the d_1, $\bar{\beta}$ plane. However, two lines for $F_i^* = 0$ appear in this plane, and we must decide which one will be effective in altering the motion. This fact is easily determined since F_1^* was initially greater than zero. The effective switching plane is the one which is approached by points coming from $F_1^* > 0$, presented by the line in the d_1, β plane limiting the region with $F_1^* > 0$; in this case, the right hand one. The easiest way to construct the projection in the d_1, $\bar{\beta}$ plane is simply to indicate several pertinent points and sketch the remainder of the curve. The fact that the projection of the phase curve in this plane is a damped cosine line is an invaluable aid in drawing the curve accurately and rapidly.[1]

Assuming $\nu\tau = \pi/8$, we may indicate the point T_1 on the spiral in the χ, χ' plane. Equation (94b)—written for control system C_1—gives the value of $\bar{\beta}$ at the point T_1 as

$$\bar{\beta} = \bar{\beta}_i - (\operatorname{sgn} F_1^*) \frac{1}{\nu} \nu\tau$$

Then $d_1(T_1)$ and $\bar{\beta}(T_1)$ will enable us to plot this point in the d_1, $\bar{\beta}$ plane. A line connecting P_i and T_1 will be a portion of a damped cosine line with its neutral axis at $2D\sqrt{1 - D^2}/\sqrt{1 - 2\kappa D + \kappa^2}$ according to equation (104b). Note that the point T_1 lies on the same side of the switching plane as does P_i; therefore, the motion must be continued until a switch point has been reached.

Choosing $\nu\tau = \pi/4$, by analogous construction we find the point T_2, which in the d_1, $\bar{\beta}$ plane lies on the opposite side of $F_1^* = 0$ from P_i. It is obvious that the damped cosine line cuts the switching plane at the point S_{1L} lying between T_1 and T_2. The value of d_{1L} will then enable us to find the point S_{1L} in the χ, χ' plane.[2]

Now, it must be remembered that χ' and d_1 (see equation 104a) change discontinuously at every switch point because $\bar{\beta}'$ is discontinuous at switch points (see equation (95)). Therefore, in the d_1, $\bar{\beta}$ plane we must go from S_{1L} to S_{1R} on the other switching plane $F_1^* = 0$ as a limit of $F_1^* < 0$. It is

[1] For accurate construction of cosine lines it is suggested that one adds to the selected points those which are essential for its form: those where $|d_1|$ has its greatest value or where the neutral line is crossed.

[2] The subscripts L and R designate those points which, in a time-dependent diagram, would lie just to the left or right-hand sides of the switch points.

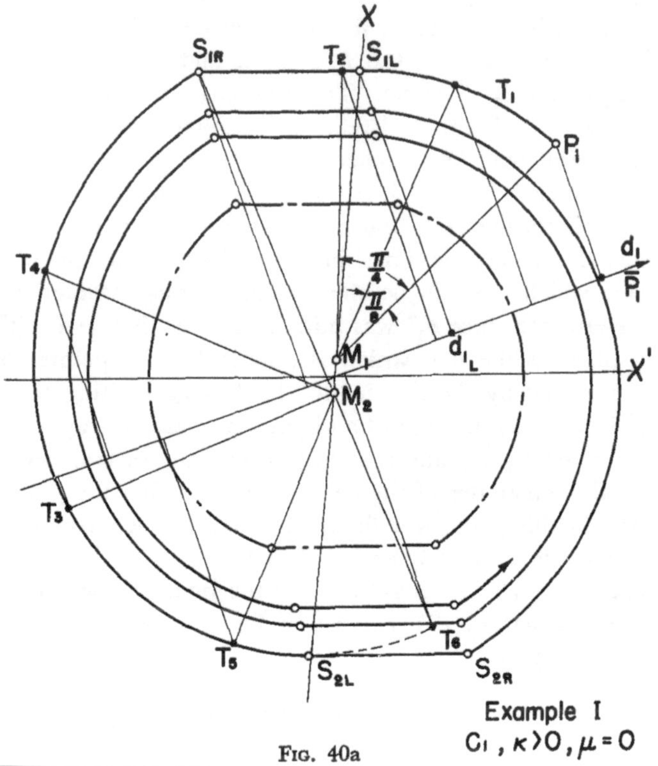

FIG. 40a

Table III[1]

M	$\bar{\beta}_i$	sgn F_1^*	ν_T	$\bar{\beta}$	T
M_1	-0.7	$+1$	$\frac{\pi}{8}$	-1.09	T_1
			$\frac{\pi}{4}$	-1.48	T_2
M_2	-1.44	-1	$\frac{\pi}{2}$	$+0.13$	T_3
			$\frac{\pi}{4}$	-0.66	T_4
			$\frac{3\pi}{4}$	$+0.92$	T_5
			π	$+1.70$	T_6
M_1	$+1.23$	$+1$	etc.		

[1] Note that $\tau = 0$ at P_i, and that at every switch point the new interval is started with $\tau = 0$.

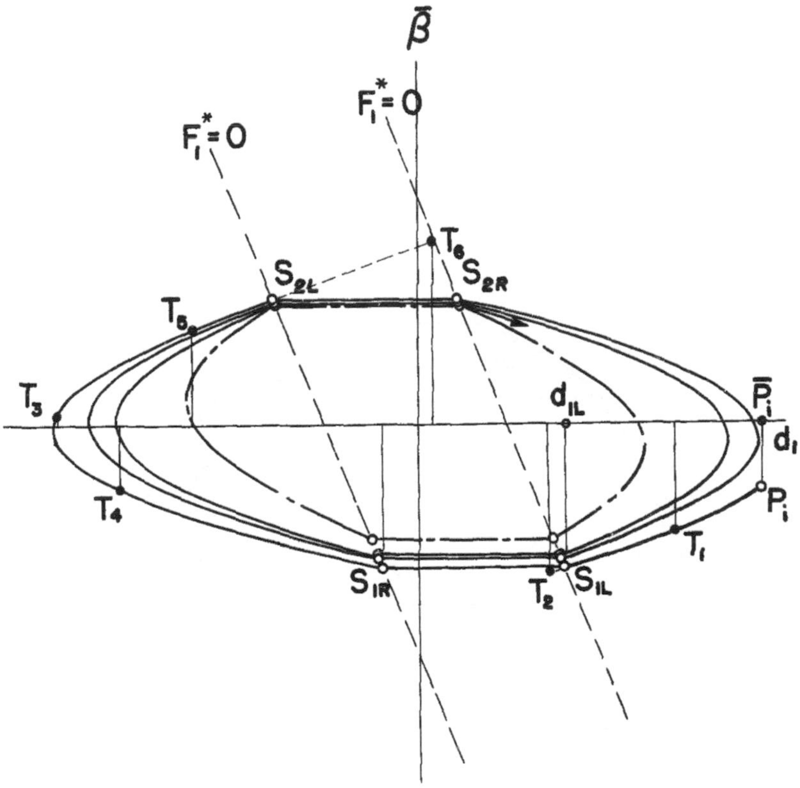

FIG. 40b

required for a continuous function that $\bar{\beta}_{1L} = \bar{\beta}_{1R}$. The coordinate d_{1R} aids in locating S_{1R} in the χ, χ' plane, where $\chi_{1L} = \chi_{1R}$ owing to the continuity of $\chi = \bar{\psi} - \bar{\beta}$. As a check on the foregoing, one may use the expression

$$|\chi'_{1R} - \chi'_{1L}| = 2, \tag{105}$$

an equality corresponding to equations (95) and (96b).

To continue the construction, we begin anew at the point S_{1R} in the χ, χ' plane, but the spiral is generated around the center $M_2 = -2D$. We look for the next switch point S_{2L}. Here again we must observe the discontinuity of d_1 and χ'. In the same manner we progress to subsequent switch points for as many convolutions as are desired.

Table III shows the various "trials," that is, various values of ν_T which were selected for the construction of this example. The motion tends toward a periodic one which is shown in the Figs. 40a, b, by a dashed line.

5·2. Different types of motion (an introduction to the complete study). Now that the method of representation has been established, we may, as for the case of position control, present a few examples of this type of control.

Example II. A motion is shown in Figs. 41a, b, which tends towards the same periodic motion as Example I did. However, the motion in the d_1,

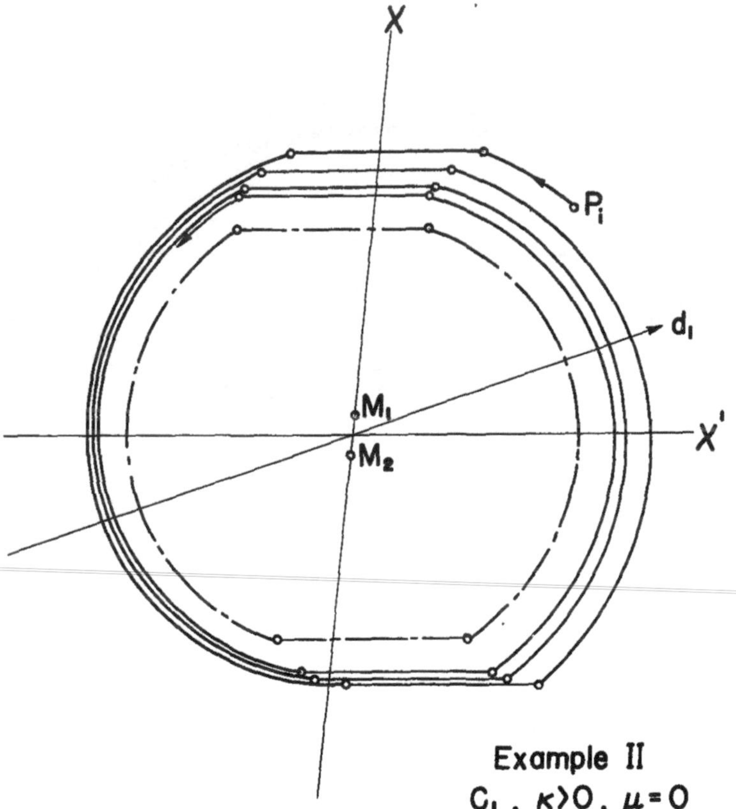

FIG. 41a

$\bar{\beta}$ plane looks quite different. This is easily explained as the result of selecting the large negative value of $\bar{\beta}$ initially.

Because the first two examples utilized system C_1, we now choose one of system C_2 as *Example III.* As may be seen from Figs. 42a, b, the construction frame of the χ, χ' plane looks much the same as for the two examples of system C_1. However, the frame in the d_1, $\bar{\beta}$ plane is markedly different from the previous examples because of the system change. The two half spaces representing $F_1^* > 0$ and $F_1^* < 0$ are not separated by a gap, as previously, but are overlapping (see sketch in Fig. 42b). The projection of the phase curve in the d_1, $\bar{\beta}$ plane must thus have a new form.

Turning our attention to the actual phase curve projections in Figs.

42a, b, we note that upon leaving the starting point P_i the switch point $S_{1L}(S_{1R})$ occurs. Past this point the switch plane will not be cut again. The spiral in the χ, χ' plane is turning about M_1, which means that

$$\chi_\infty \to +2D$$
$$\chi'_\infty \to 0$$

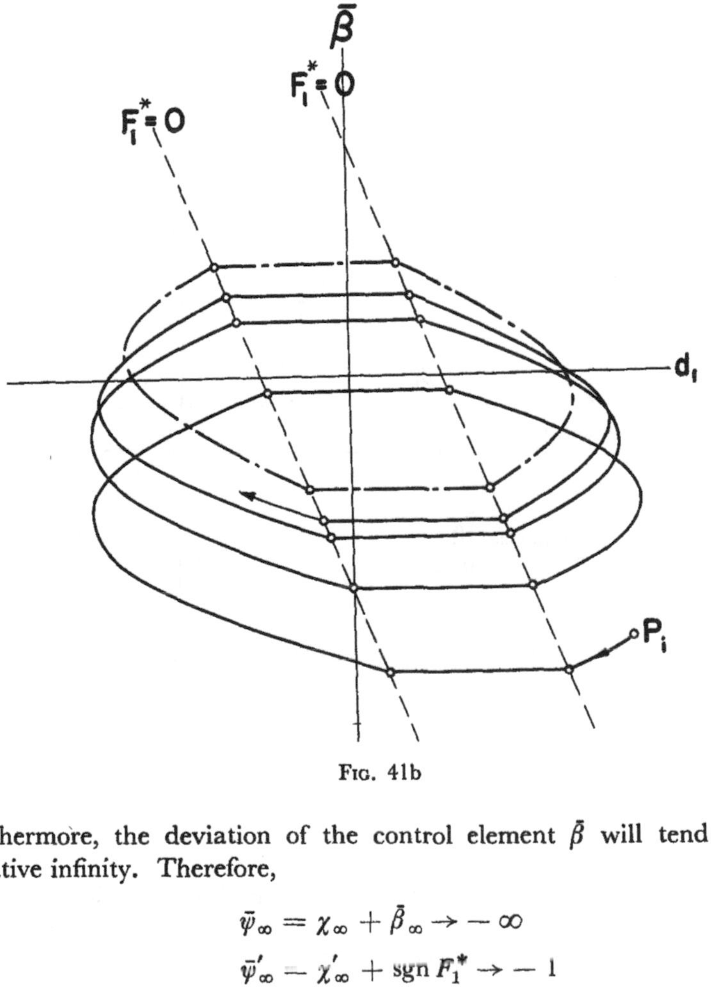

FIG. 41b

Furthermore, the deviation of the control element $\bar{\beta}$ will tend towards negative infinity. Therefore,

$$\bar{\psi}_\infty = \chi_\infty + \bar{\beta}_\infty \to -\infty$$
$$\bar{\psi}'_\infty - \chi'_\infty + \operatorname{sgn} F_1^* \to -1$$

Thus "control at rest" in a system with discontinuous velocity control means a steadily growing absolute deviation.

Let us observe the effects of feedback on motions with velocity control. *Example IV*, Figs. 43a, b, depicts a motion for system C_1, $\kappa = 2.3$, and $\mu = +1$. Point S_{3R} does not allow real "switching." The motion becomes undefined for an ideal control mechanism. But in the practical case the existence of time lags will initiate a motion of very high frequency tending towards $\chi = 0$, $\chi' = + \operatorname{sgn} F_1^*$, $\bar{\beta} = 0$ (see Figs. 44a, b). In other words $\bar{\psi}$, $\bar{\psi}'$ and $\bar{\beta}$ all tend towards zero.

After such an end point the average motion due to a small time lag may be found by taking the average value of F_1^* to be zero.

$$F_1^* = \bar{\psi} + \kappa\bar{\psi}' + \mu\bar{\beta} = 0 \tag{106}$$

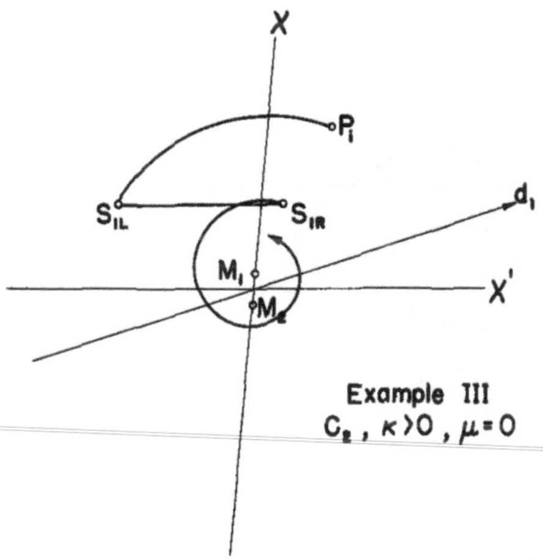

FIG. 42a

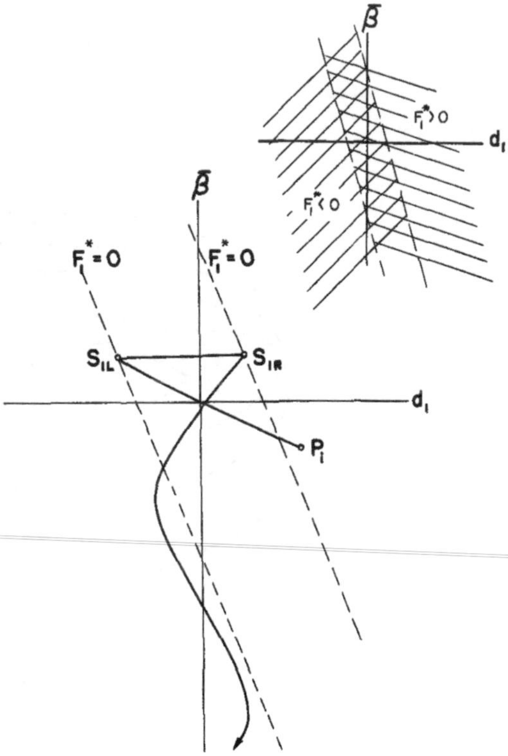

FIG. 42b

Upon combining this equation with the equation of motion (86a)

$$\bar{\psi}'' + 2D\bar{\psi}' + \bar{\psi} = \bar{\beta} \tag{107}$$

we obtain the following for the average motion:

$$\bar{\psi}_{\text{av}}'' + \bar{\psi}_{\text{av}}'\left(2D + \frac{\kappa}{\mu}\right) + \bar{\psi}_{\text{av}}\left(1 + \frac{1}{\mu}\right) = 0 \tag{108}$$

In Example IV $\mu > 0$ and $\kappa/\mu > 0$, hence the functions $\bar{\psi}$, $\bar{\psi}'$, and $\bar{\beta}$ will tend towards zero rapidly, a result in agreement with that deduced from

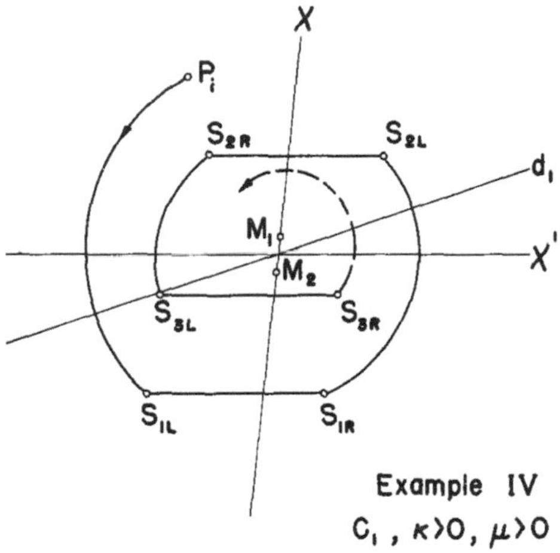

FIG. 43a

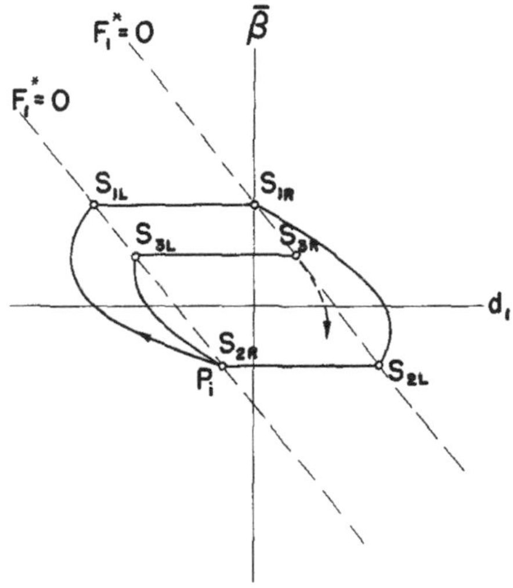

FIG. 43b

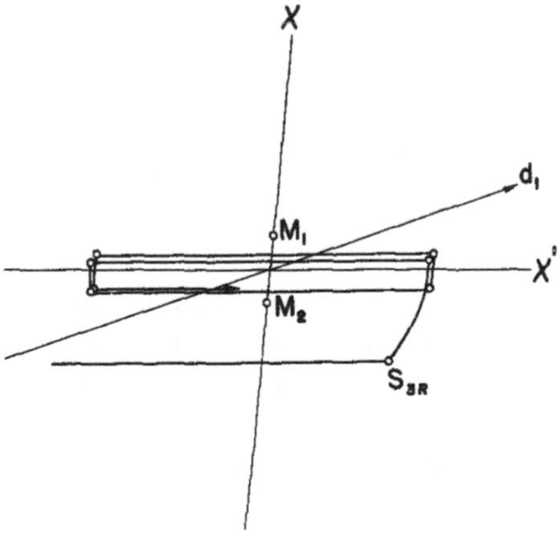

FIG. 44a

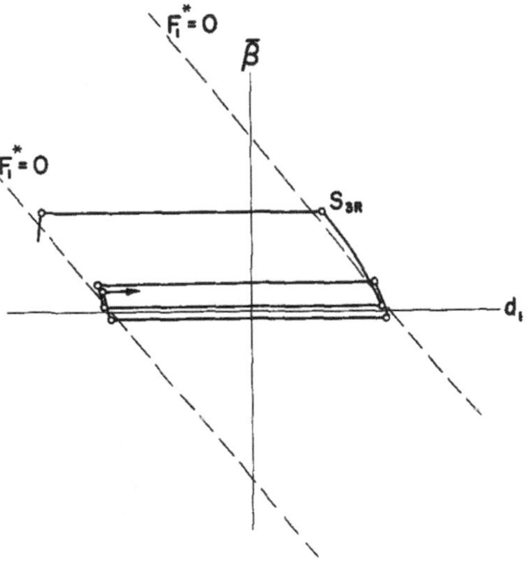

FIG. 44b

the construction of the phase curve. We may conclude the numerical analysis with the following data:

$$\kappa = 2.3 \text{ and } \mu = 1$$

$$1 + \frac{1}{\mu} = 2$$

$$2D + \frac{\kappa}{\mu} = 2.5$$

$$\bar{\psi}_{av} = 0.245\, e^{-1.25\tau} \cos(0.663\tau + 1.012)$$

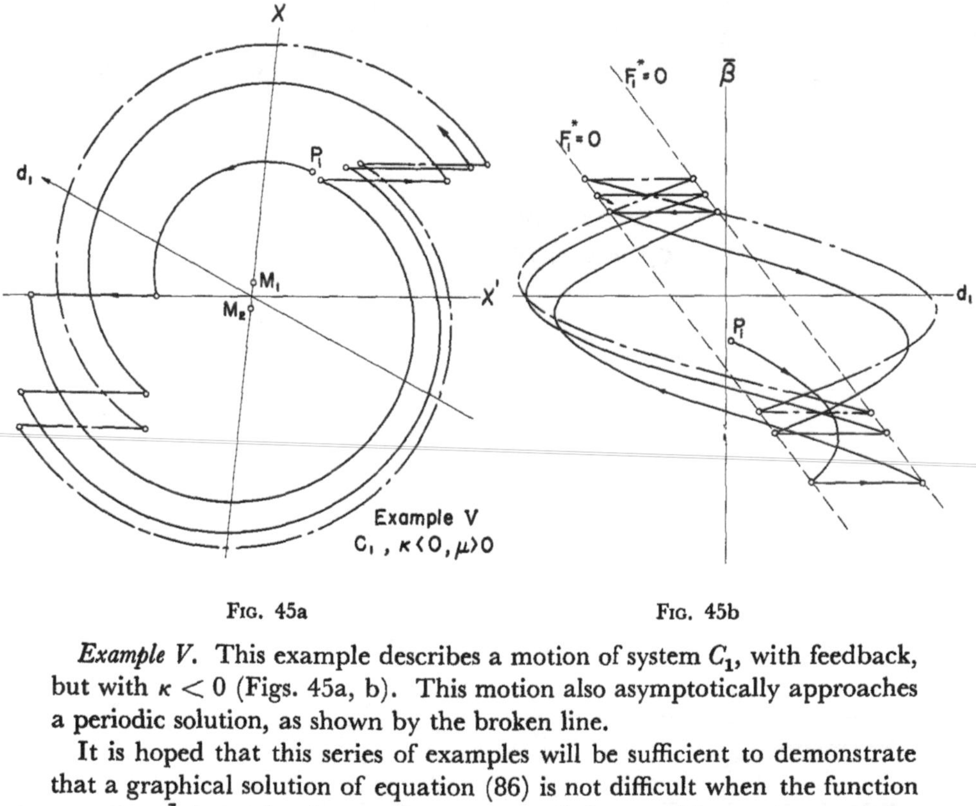

FIG. 45a FIG. 45b

Example V. This example describes a motion of system C_1, with feedback, but with $\kappa < 0$ (Figs. 45a, b). This motion also asymptotically approaches a periodic solution, as shown by the broken line.

It is hoped that this series of examples will be sufficient to demonstrate that a graphical solution of equation (86) is not difficult when the function $\chi = \bar{\psi} - \bar{\beta}$ is used. A complete survey of the possible motions will be found in section (4) of this chapter.

5·3. The control function F_1^* as a function of χ, χ', and $\bar{\beta}$. The control function F_1^* is given by

$$F_1^* = \chi + \kappa\chi' + \bar{\beta}(1 + \mu) \mp \kappa \operatorname{sgn} F_1^* \tag{97}$$

or, combined with equation (104a),

$$F_1^* = \frac{\sqrt{1 - 2\kappa D + \kappa^2}}{\sqrt{1 - D^2}}\, d_1 + \bar{\beta}(1 + \mu) \mp \kappa \operatorname{sgn} F_1^* \tag{109}$$

$F_1^* = 0$ defines two planes in the χ, χ', $\bar{\beta}$ space as indicated earlier. In the d_1, $\bar{\beta}$ plane they appear as straight lines symmetrical to the origin.

The distance of any point $(d_1, \bar{\beta})$ from the lines $F_1^* = 0$ in the d_1, $\bar{\beta}$ plane is proportional to the value of F_1^*. This fact enables one to determine the sign of F_1^* immediately (see Fig. 46). The extreme values of F_1^* for a given phase curve are found from the condition that

$$\frac{dF_1^*}{d\tau} = 0$$

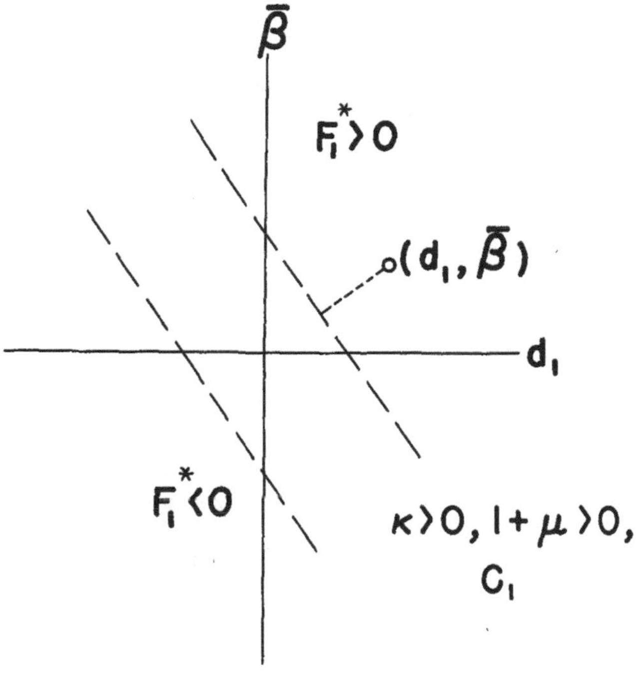

FIG. 46

From equation (87) we immediately find that

$$F_1^{*\prime} = \bar{\psi}' + \kappa\bar{\psi}'' + \mu\bar{\beta}' \tag{110}$$

and by introducing equation (86), we obtain

$$\begin{aligned} F_1^{*\prime} &= \bar{\psi}' + \kappa(-2D\bar{\psi}' - \bar{\psi} + \bar{\beta}) + \mu(\mp \operatorname{sgn} F_1^*) \\ &= \bar{\psi}'(1 - 2\kappa D) - \kappa(\bar{\psi} - \bar{\beta}) + \mu(\mp \operatorname{sgn} F_1^*) \end{aligned} \tag{111}$$

Now the derivative of the control function may be written in terms of the χ, χ' system (by equation (91))

$$\begin{aligned} F_1^{*\prime} &= (\chi' \mp \operatorname{sgn} F_1^*)\,(1 - 2\kappa D) - \kappa\chi + \mu(\mp \operatorname{sgn} F_1^*) \\ &= \chi'\,(1 - 2\kappa D) - \kappa\chi \mp \operatorname{sgn} F_1^*\,(1 - 2\kappa D + \mu) \end{aligned} \tag{112}$$

Because this equation does not contain $\bar{\beta}$, the interesting result occurs that all points $F_1^{*\prime} = 0$ lie on two planes normal to the χ, χ' plane. The lines of intersection of these planes with the χ, χ' plane should always be drawn as part of the basic frame (Sec. 5.4) because they facilitate the construction by indicating the extreme values of the function F_1^*.

Equation (112) clearly shows that $F_1^{*\prime}$ is discontinuous at switch points. Although the terms χ and $(\chi' \mp \operatorname{sgn} F_1^*)$ are continuous, the term $\mu(\mp \operatorname{sgn} F_1^*)$ will change discontinuously as long as $\mu \neq 0$, that is, as long as the control operates with feedback.

The intersection lines of the $F_1^{*\prime} = 0$ planes and the χ, χ' plane depend upon μ and κ. It can be shown that for constant μ the intersection lines for various κ all go through a fixed point. Differentiating equation (112) with respect to κ, we obtain

$$\frac{dF_1^{*\prime}}{d\kappa} = (\chi' \mp \operatorname{sgn} F_1^*)\,(-2D) - \chi = 0 \tag{113a}$$

which together with

$$F_1^{*\prime} = (\chi' \mp \operatorname{sgn} F_1^*)\,(1 - 2D\kappa) - \kappa\chi \mp \mu \operatorname{sgn} F_1^* = 0 \tag{113b}$$

determines the fixed point:

$$\left.\begin{aligned} \chi &= \mp 2D\mu \operatorname{sgn} F_1^* \\ \chi' &= -(1 + \mu)\,(\mp \operatorname{sgn} F_1^*) \end{aligned}\right\} \tag{114a}$$

To aid in drawing the line $F_1^{*\prime} = 0$ one may also use the point given by

$$\left.\begin{aligned} \chi' \mp \operatorname{sgn} F_1^* &= 0 \\ \chi = \mp (\mu/\kappa) &\operatorname{sgn} F_1^* \end{aligned}\right\} \tag{114b}$$

in addition to the fixed point.

The planes given by $F_1^{*\prime} = 0$, which are perpendicular to the χ, χ' plane, cut the corresponding switching planes $F_1^* = 0$ in two intersection lines. $F_1^* = 0$ and $F_1^{*\prime} = 0$ coincide exactly at all points of these intersection lines. Therefore these points will be of importance in aiding us to determine rest, end, and starting points for a given control system.

5.4. Complete survey of the possible motions. (1) *The construction of the "frame" and its dependence on κ and μ; length of intervals.* The choice of the system (C_1 or C_2) and the chosen values of κ and μ will determine the motion a body with velocity control will have.

In order to present this survey in the simplest and most logical manner the concept of the construction frame, already referred to, must be developed. Holding the parameter D fixed, we proceed to vary the coefficients κ and μ in system C_1 or C_2. In the χ, χ' plane the frame consists of the χ, χ', and d_1 axes, and the lines $F_1^{*\prime} = 0$. In the $\bar{\beta}$, d_1 plane the frame consists of the d_1

and $\bar{\beta}$ axes and the switch lines $F_1^* = 0$. Figs. 47a-h show these frames for all possible types of motion. Phase curves have been shown for illustration, and wherever possible a periodic curve has been taken for the sake of simplicity.

In Fig. 47b segments of the spirals are marked corresponding to their centers; the lines $F_1^{*\prime} = 0$ are marked in a similar manner (see equation (112)). To construct figures one should note the corresponding portions of the motion by the appropriate center. Generally this procedure has not been followed in order to avoid complicating the figures.

The angle σ between the χ and χ' axes (Fig. 34) is given by the damping coefficient: $\cos \sigma = -D$. The lines $F_1^{*\prime} = 0$ in the χ, χ' plane are determined by equation (113b); their construction is described in the preceding section. The switching lines in the d_1, $\bar{\beta}$ plane are given by equation (109) and are located with the aid of the expressions for $\bar{\beta}_{00}$ and d_{00} (equations (98) and (101)). Because the signs of these two segments depend upon κ and $(1 + \mu)$, it is clear that the frame will be greatly influenced by the parameters κ and $(1 + \mu)$ as well as by the choice of the system. For $(1 + \mu) = 0$ the switching lines will be normal to the d_1 axis.

The following figures all have very similar frames in the d_1, $\bar{\beta}$ plane because the two half spaces, $F_1^* > 0$ and $F_1^* < 0$, are separated by a gap:

Fig.	κ	*System*	$(1 + \mu)$
47a	$\kappa > 0$	C_1	$(1 + \mu) > 0$
47b	$\kappa < 0$	C_2	
47c	$\kappa > 0$	C_1	$(1 + \mu) < 0$
47d	$\kappa < 0$	C_2	

The remaining figures of the series are also very similar because of their appearance in the d_1, $\bar{\beta}$ plane. In this case the two half spaces overlap:

Fig.	κ	*System*	$(1 + \mu)$
47e	$\kappa < 0$	C_1	$(1 + \mu) > 0$
47f	$\kappa > 0$	C_2	
47g	$\kappa < 0$	C_1	$(1 + \mu) < 0$
47h	$\kappa > 0$	C_2	

The direction of motion in the χ, χ' plane is fixed, for the spirals have only one sense of motion. In the d_1, $\bar{\beta}$ plane the sense of motion is given by

$$\bar{\beta}_m = \bar{\beta}_{mi} \mp (\operatorname{sgn} F_1^*)\tau_m \qquad (94)$$

where $\tau_m > 0$. For example in the half space $F_1^* > 0$ this equation requires that $\bar{\beta}$ will decrease for system C_1 and will increase for system C_2.

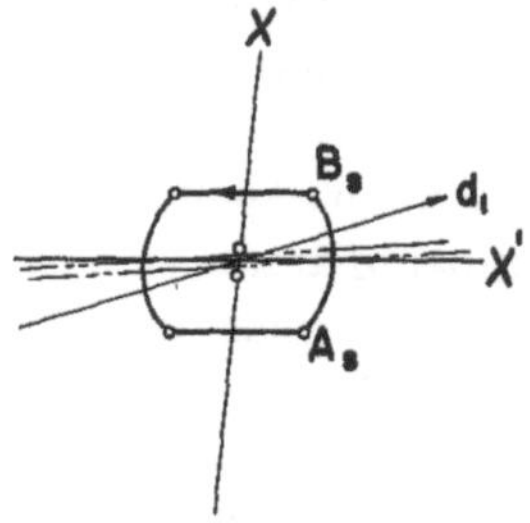

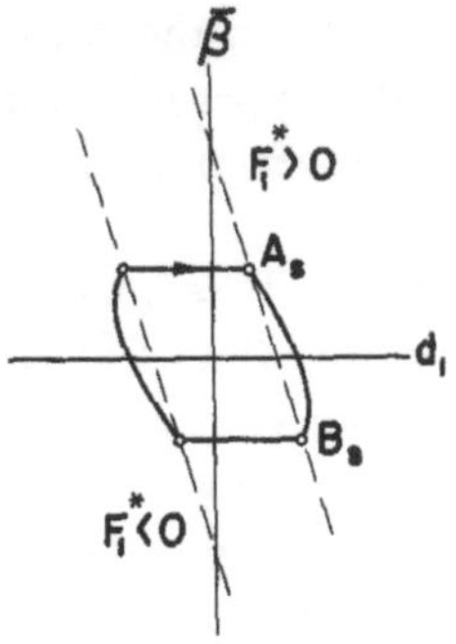

$1+\mu > 0$
$\kappa > 0$, C_1
Fig. 47a

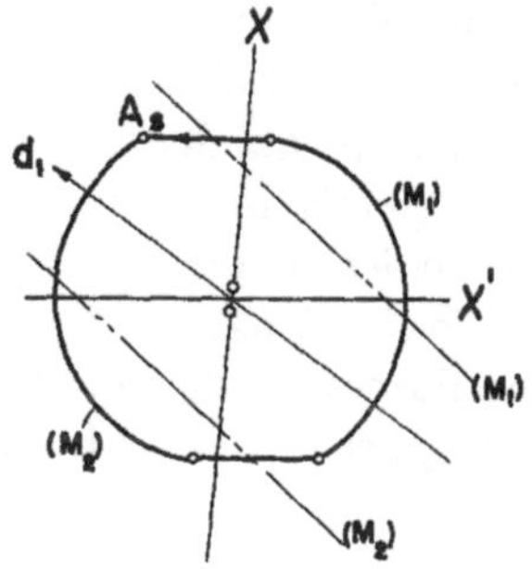

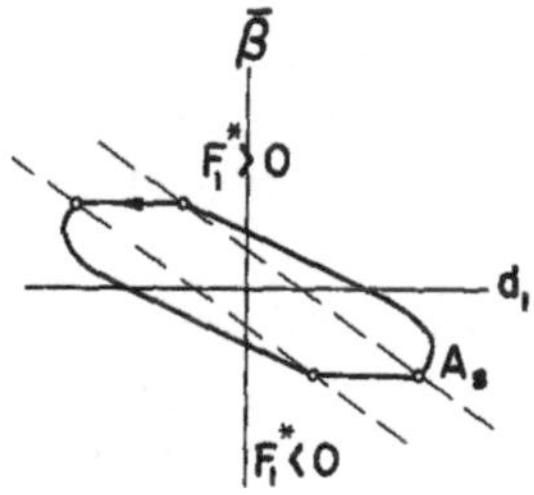

$1+\mu > 0$
$\kappa < 0$, C_2
Fig. 47b

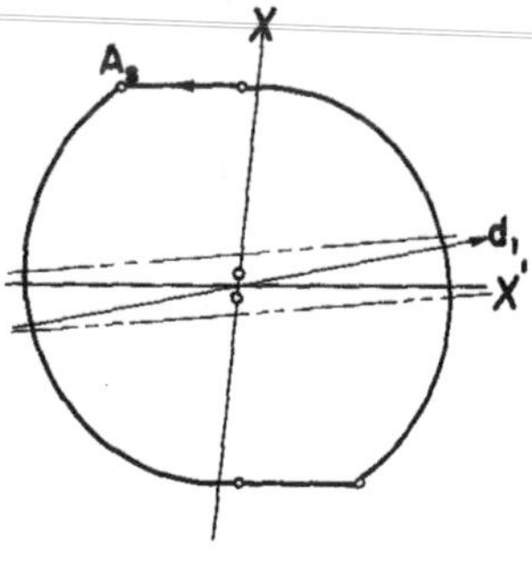

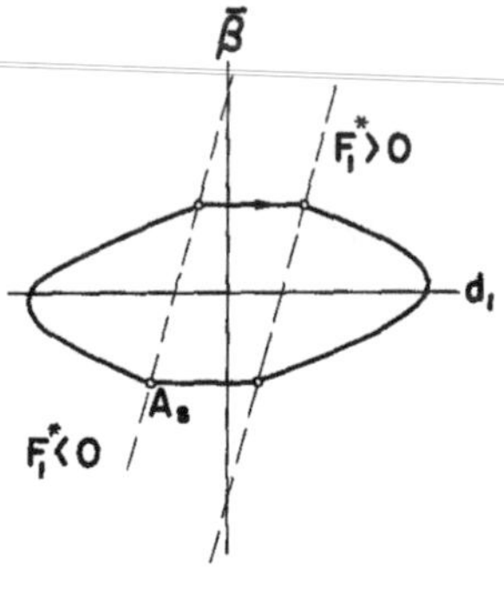

$1+\mu < 0$
$\kappa > 0$, C_1
Fig. 47c

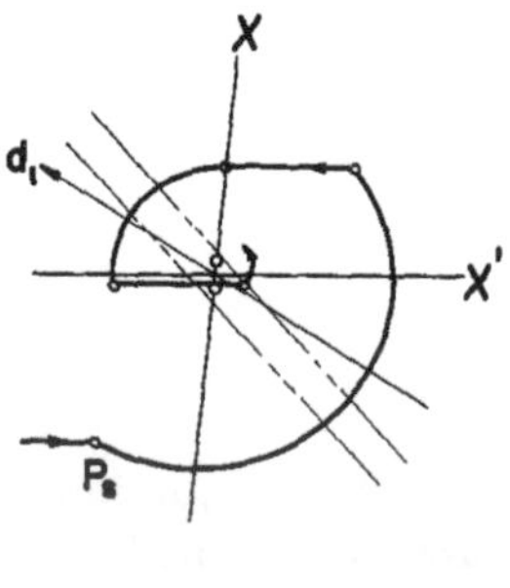

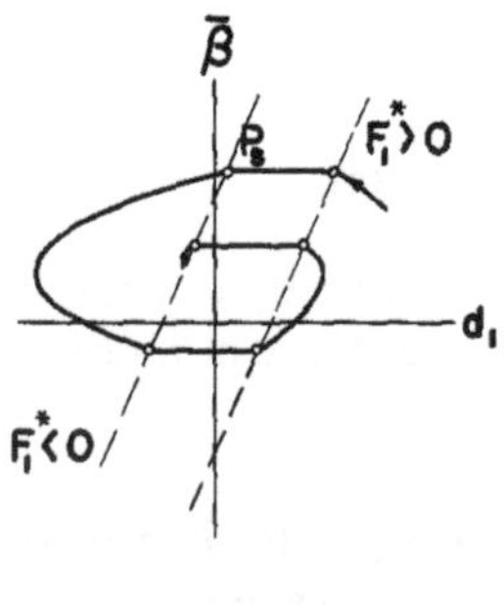

$1+\mu < 0$
$\kappa < 0$, C_2
Fig. 47d

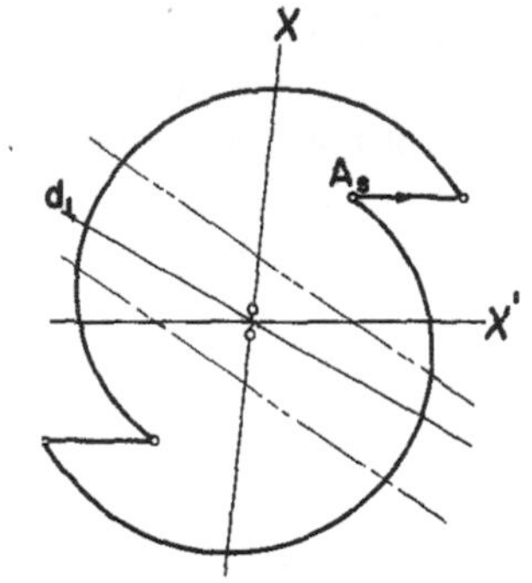

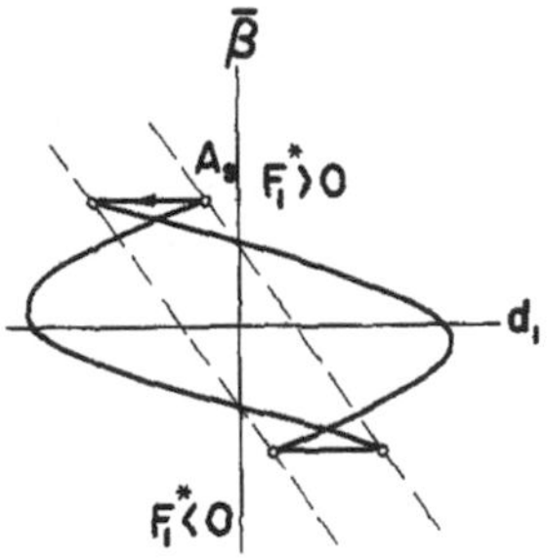

$1+\mu > 0$
$\kappa < 0$, C_1
Fig. 47e

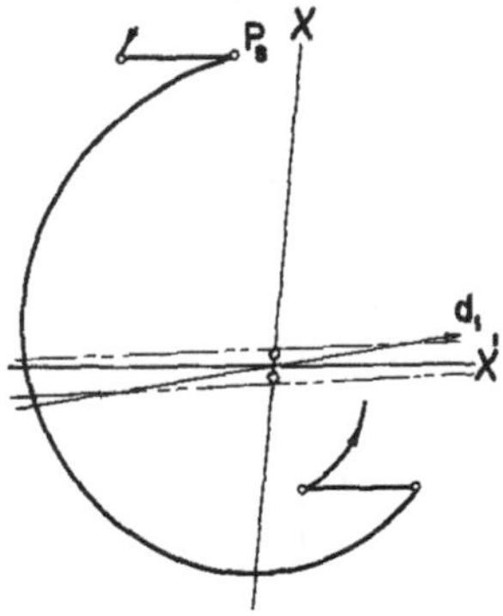

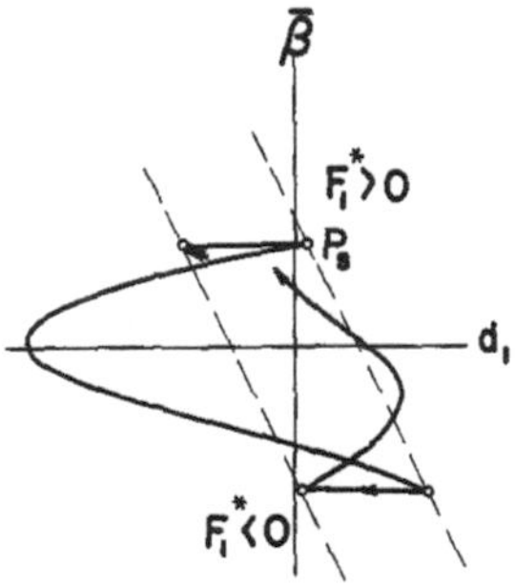

$1+\mu > 0$
$\kappa > 0$, C_2
Fig. 47f

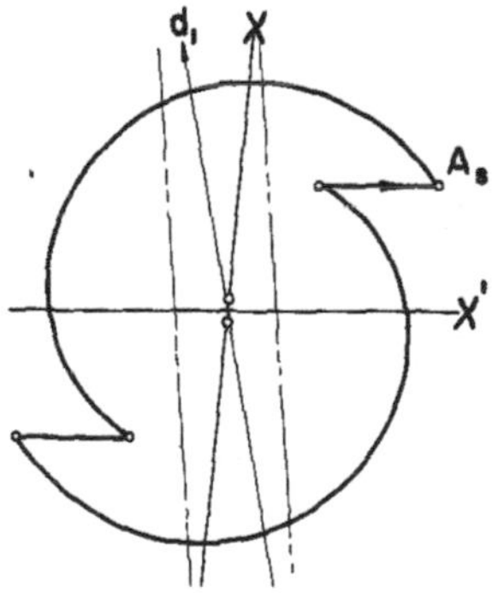

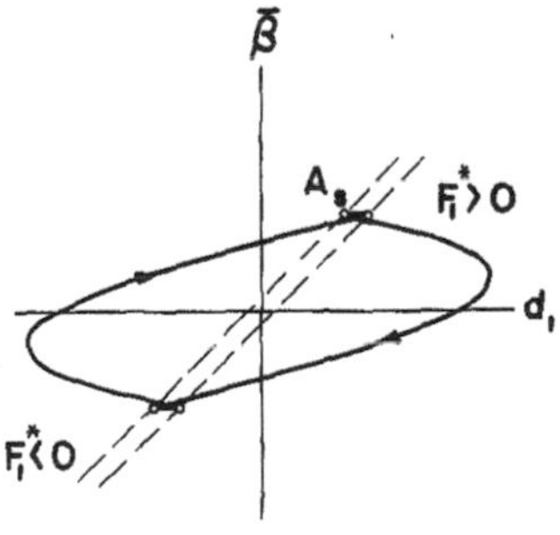

$1+\mu < 0$
$\kappa < 0$, C_1
Fig. 47g

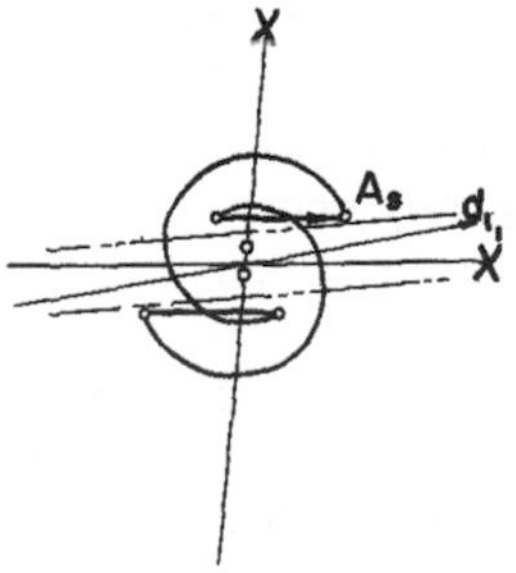

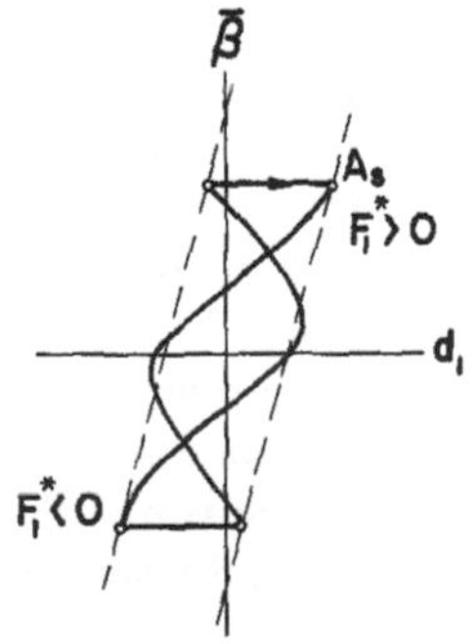

$1+\mu < 0$
$\kappa > 0$, C_2
Fig. 47h

Figs. 47a-h also serve to give some ideas regarding the lengths of the intervals which occur. In the position control problem the interval length could be judged readily by observing the location of the spiral center in the phase plane (Sec. 4.5). However, in the velocity control problem it is obvious that the length of intervals is not influenced so much by the location

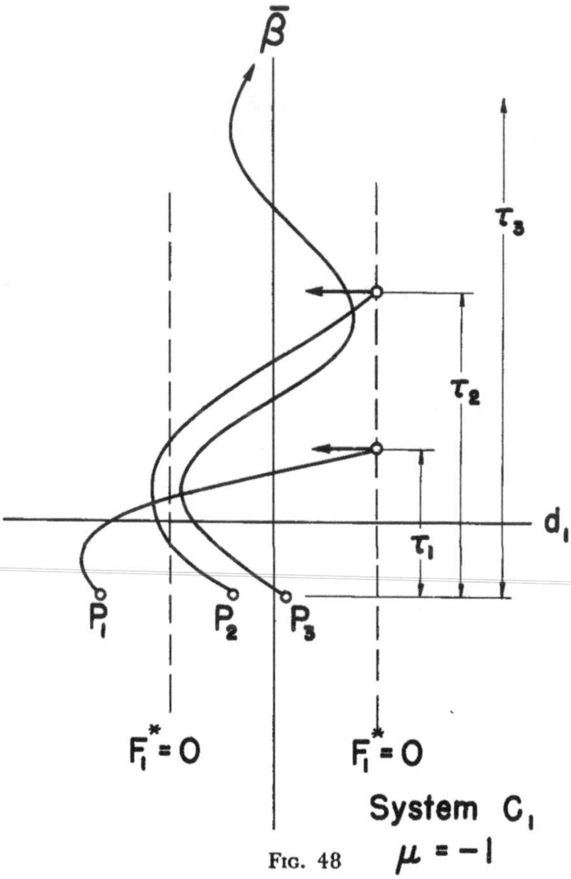

FIG. 48

of M_1 and M_2 (they are much closer together than for position control: $|\overline{M_1M_2}| = 4D$) as by the possibility that the two half spaces overlap or are separated by a gap. In general, it appears that the interval lengths in frames with overlapping half spaces will be larger than those in frames with separated half spaces. This overlapping may occur in either system C_1 or C_2 and because rest of control may occur in either system there is actually no upper limit to the interval length.

The influence of the coefficients κ and μ upon the interval length may be studied conveniently by means of two examples (Figs. 48 and 49). First let us look at the special case for which $\mu = -1$ (Fig. 48). In this example the switch lines $F_1^* = 0$ are normal to the d_1 axis (an overlapping system typified by Fig. 47e is shown). The initial value of d_1 depends upon χ_i, χ_i', κ and D; this value of d_1 will determine the length of the first interval.[1] The figure shows that any length of interval up to infinity—corresponding to rest of control—is possible for this special frame.

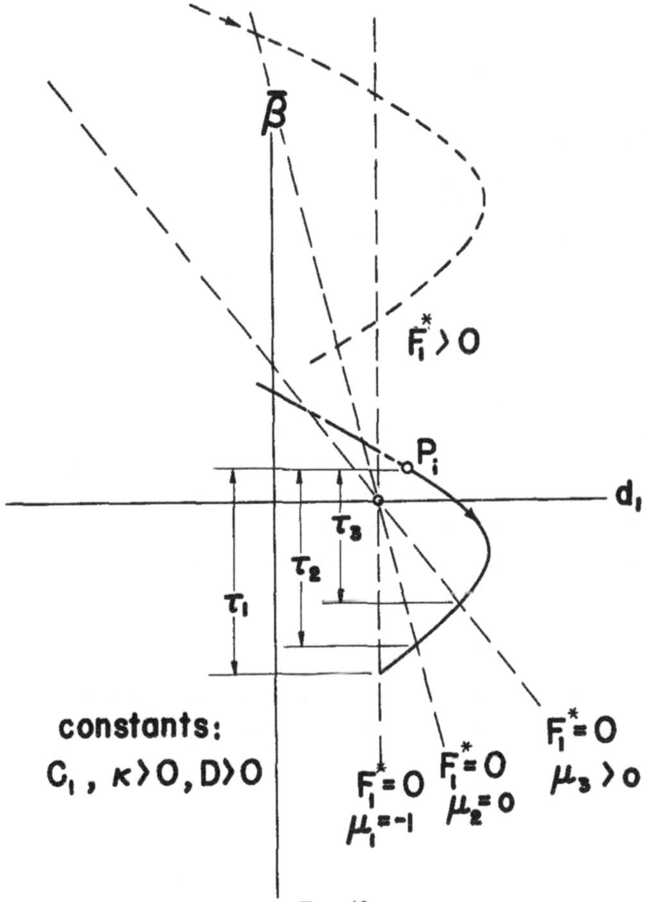

FIG. 49

Now let us *fix a portion of a spiral* in the χ, χ' plane and study the effect upon the length of the first interval when only μ is allowed to vary (for the sake of simplicity we will choose an example of the non-overlapping type). This set of conditions shows that the sole effect of varying μ will be to change δ, which determines the slope of the switch lines in the d_1, $\bar{\beta}$ plane. With $\mu \neq -1$ (the switch lines will then be inclined to the d_1 axis) we see in Fig. 49 that as μ attains greater positive values the length of the first interval

[1]The "first interval" is defined as that portion of the motion occurring between the start of the motion and the first switch point.

decreases. Because a portion of a spiral in the χ, χ' plane is fixed and κ is constant, all curves in Fig. 49 for various μ are portions of the same damped cosine line in the d_1, $\bar{\beta}$ plane. Although not shown, it should be evident that for $\mu < -1$ the length of interval would increase.

Considering P_i not as a starting point but as a point which is passed through during the course of a motion, we might use Fig. 49 for estimating the length of any interval between two switch points. It is obvious that the length of these intervals depend upon κ and μ; however, the average value of $\bar{\beta}$ has an even greater influence. For example, observe the dashed projection of a phase curve in the $\bar{\beta}$, d_1 plane. This curve has the same χ, χ' projection as the former one. Of course, these examples do not give a complete survey of the question of interval length, but they should serve as a useful guide for further discussion.

The known results follow: (1) if rest points exist, the maximum interval length is infinite; (2) if the motion tends towards a periodic motion, then the interval length will tend toward a particular value; and (3) if the motion tends towards an end point, passing through this state because of imperfections of the system, periodic motions with very short interval lengths may result (Fig. 44 a/b).

(2) *Periodic motions.* The conditions for periodicity are relatively easy to establish if one determines switch point conditions for a periodic motion. It is evident that the period must consist of two parts: one part of the phase curve must be contained in the region $F_1^* > 0$ and the other in the region $F_1^* < 0$. Therefore three consecutive switch points must be considered. Periodicity requires that

$$\chi_1 = \chi_3, \quad \chi'_1 = \chi'_3, \quad \bar{\beta}_1 = \bar{\beta}_3 \tag{115}$$

Let us assume for the moment that the first part of the period has the length τ_{p1}, and the second part, the length τ_{p2}. Then we have

$$\left.\begin{aligned} \bar{\beta}_2 &= \bar{\beta}_1 \mp (\operatorname{sgn} F_1^*)_1 \cdot \tau_{p1} \\ \text{and} \qquad \bar{\beta}_3 &= \bar{\beta}_2 \mp (\operatorname{sgn} F_1^*)_2 \cdot \tau_{p2} \end{aligned}\right\} \tag{116}$$

Since $(\operatorname{sgn} F_1^*)_1 = -(\operatorname{sgn} F_1^*)_2$, we may write that

$$\left.\begin{aligned} \bar{\beta}_2 &= \bar{\beta}_1 \mp (\operatorname{sgn} F_1^*)_1 \cdot \tau_{p1} \\ \text{and} \qquad \bar{\beta}_3 &= \bar{\beta}_2 \pm (\operatorname{sgn} F_1^*)_1 \cdot \tau_{p2} \end{aligned}\right\} \tag{117}$$

Therefore,

$$\bar{\beta}_3 = \bar{\beta}_1 \mp (\operatorname{sgn} F_1^*)_1 \cdot (\tau_{p1} - \tau_{p2}) \tag{118}$$

Equation (118) will satisfy equation (115) only if $\tau_{p1} = \tau_{p2}$, which indicates that the period consists of two parts of equal length.

In the χ, χ' plane the two parts of the periodic phase curve must be situated correspondingly to M_1 and M_2 in order that $\tau_{p1} = \tau_{p2}$. In turn this fact requires that $d_{12} = -d_{11}$, and this further leads to $\bar{\beta}_2 = -\bar{\beta}_1$ (note any one of the periodic motions depicted in Figs. 47).

The conditions of periodicity given in equation (115) may now be extended to include the intermediate switch point.

$$\left.\begin{aligned} \chi_1 &= \chi_3 = -\chi_2 \\ \chi'_{1R} &= \chi'_{3R} = -\chi'_{2R} \\ \bar{\beta}_1 &= \bar{\beta}_3 \ \ = -\bar{\beta}_2 \end{aligned}\right\} \tag{119}$$

The discontinuity of χ' at the switch points requires that conditions be set for either χ'_{iR} or χ'_{iL} (see Fig. 40 and footnote 3 of this chapter for the meaning of the subscripts R and L).

Equations (119) may be converted into analytical form by means of equations (91) and (94b). A tedious elementary computation leads to the following formulas:

$$\bar{\beta}_p = \pm \frac{1}{2} \operatorname{sgn} F_1^* \cdot \left(\frac{1}{\nu}\right) \cdot (\nu\tau_p) \tag{120a}$$

$$\chi_p = (\mp \operatorname{sgn} F_1^*) \frac{2D \sinh D\tau_p + \dfrac{1 - 2D^2}{\sqrt{1 - D^2}} \sin \nu\tau_p}{\cosh D\tau_p + \cos \nu\tau_p} \tag{120b}$$

$$\chi'_p \mp \operatorname{sgn} F_1^* = \bar{\varphi}'_p = (\mp \operatorname{sgn} F_1^*) \frac{\dfrac{D}{\sqrt{1 - D^2}} \sin \nu\tau_p - \sinh D\tau_p}{\cosh D\tau_p + \cos \nu\tau_p} \tag{120c}$$

At the switch points F_1^* is zero, but $\operatorname{sgn} F_1^*$ jumps from $+1$ to -1, or vice versa. In a continuous motion the switch point occurs at the end (or start) of a succession of points for which the value of $\operatorname{sgn} F_1^*$ is known. Thus to construct a periodic motion, one locates the coordinates of the switch point in the projection planes, assuming that $(\mp \operatorname{sgn} F_1^*)$ is either $+1$ or -1, and draws the spirals in the χ, χ' plane between the switch points around the center indicated by equation (92).

In order to construct the projections of this phase curve in the d_1, $\bar{\beta}$ plane, the damped cosine lines, we must know the values of κ and μ. The fact that $F_1^* = 0$ at the switch points enables us to compute the value of κ_p at these points (see equation (97)) to be

$$\kappa_p = -\frac{\chi_p + \bar{\beta}_p (1 + \mu_p)}{\chi'_p \mp \operatorname{sgn} F_1^*} \tag{120d}$$

provided that a value for μ_p has been chosen.

In Fig. 50 the following functions are plotted versus $\nu\tau_p$:

$$\frac{\chi_p}{\mp \operatorname{sgn} F_1^*} = \chi_p^* \tag{121a}$$

$$\frac{\chi_p' \mp \operatorname{sgn} F_1^*}{\mp \operatorname{sgn} F^*} = \bar{\psi}'^*_p \tag{121b}$$

$$\frac{\bar{\beta}_p}{\mp \operatorname{sgn} F_1^*} = \bar{\beta}_p^* \tag{121c}$$

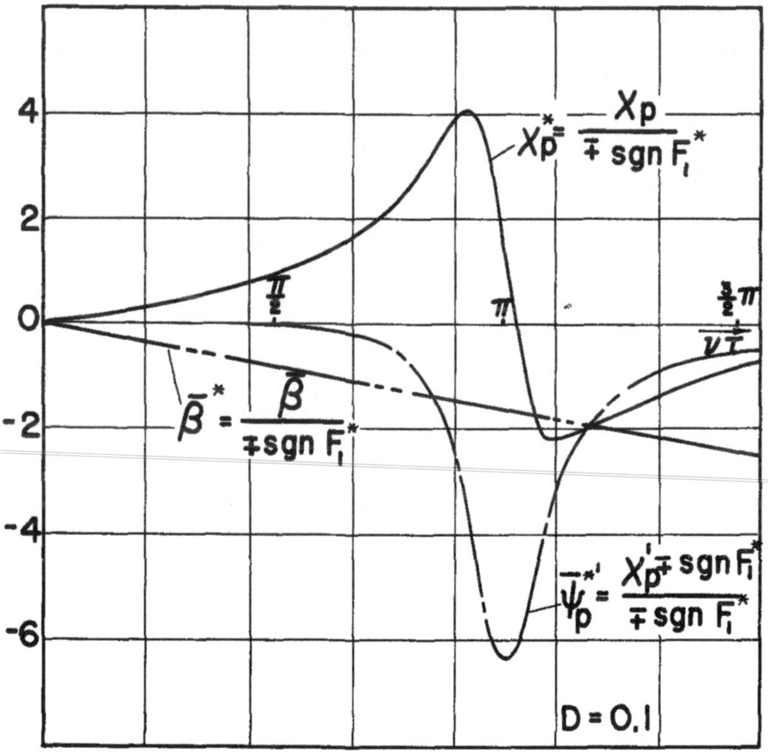

FIG. 50

[Note: $\lim_{\tau_p \to \infty} \chi_p^* = 2D$; $\lim_{\tau_p \to \infty} \bar{\psi}'^*_p = -1$]. From equation (120d), we find

$$\kappa_p = -\frac{\chi_p^* + \bar{\beta}_p^* (1 + \mu_p)}{\bar{\psi}'^*_p} \tag{122}$$

Although all combinations of κ and μ which should permit a periodic motion of the given periodic length $2\tau_p$ can be found, we cannot predict if the motion really exists and to what system (C_1 or C_2) it will belong.

Because the behavior of F_1^* between two switch points in getting equations (120) from equations (119) was not considered, we do not know that the prescribed periodic motion exists. It was assumed that F_1^* would not change its sign between switch points but *only at* those points. Now we must check the correctness of that assumption.

This problem may be studied easily by considering the relationship between F_1^* and $F_1^{*\prime}$. If F_1^* is zero at the end points of an interval, if it is continuous and different from zero throughout that interval, then $F_1^{*\prime}$ must change sign in the interval. The value of $F_1^{*\prime}$ is given by equation (112), and the lines in the χ, χ' plane where $F_1^{*\prime} = 0$ are given by equation (113b). Therefore, if the spiral between two consecutive switch points is cut only once by the lines $F_1^{*\prime} = 0$ in the χ, χ' plane, F_1^* will not change its sign between those points. However, if the line $F_1^{*\prime} = 0$ cuts the spiral more than once, the motion *may not* be possible. If the spiral is cut twice, then the function F_1^* will not satisfy the conditions stated above that the motion be the required periodic one. If the spiral is cut, say, three times, however, the conditions set forth for F_1^* may be satisfied. Such a motion would require as its half period more than one convolution of the spiral. No example of this type was met; its occurrence should be influenced largely by the size of D.

A systematic study of the existence of the periodic motions suggested by equations (120) and (121) is made in the following manner: Choose some value of τ_p and construct in the χ, χ' plane the corresponding spiral. This spiral will be independent of κ and μ. The center of the spiral M_1 or M_2 may be found by trial. A spiral of several convolutions is drawn on transparent paper and is shifted over the χ, χ' plane until the proper center is found. Equation (122) gives the possible combinations of κ and μ for a given value of τ_p, but not all of these combinations will allow a periodic motion. In a κ, μ plane equation (122) will appear as a family of straight lines with $\nu\tau_p$ as a parameter, as shown in Figs. 51a, b.

Moving systematically along such a line, taking various combinations of κ and μ we may construct the corresponding lines $F_1^{*\prime} = 0$ in the χ, χ' plane, these lines depending on both κ and μ. As long as the spiral between the two switch points is cut only once by the $F_1^{*\prime} = 0$ line, the motion is a possible one. The limiting case arises when the $F_1^{*\prime} = 0$ line passes through either of the switch points ($\nu\tau_p < 360°$). Such a limiting case is shown in Fig. 52.

Returning to Figs. 51a, b we note that these limits have been marked as broken lines. Analytically, these limits are given by combining $F_1^* = 0$ and $F_1^{*\prime} = 0$ and eliminating κ:

$$(\cosh D\tau_p - \cos \nu\tau_p) - \tfrac{1}{2}\,\tau_p \sin \nu\tau_p$$
$$+ \mu \left[\mp \frac{\dfrac{D}{\sqrt{1 - D^2}} \sin \nu\tau_p - \sinh D\tau_p}{\cosh D\tau_p + \cos \nu\tau_p} - \tfrac{1}{2}\,\tau_p \sin \nu\tau_p \right] = 0 \quad (123)$$

In general, these figures will be used in conjunction with Fig. 50 when dealing with periodic motions. Other pertinent details concerning Fig. 51 will be pointed out shortly.

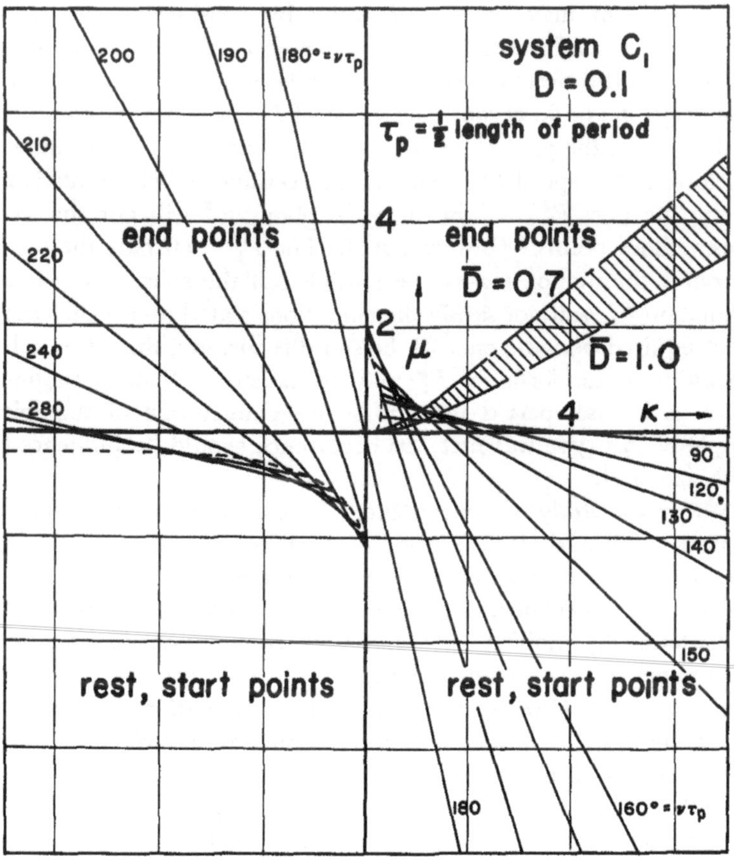

FIG. 51a

(3) *End, starting, and rest points.* Several times in the past we have indicated that the control function F_1^* changes its slope discontinuously at switch points when $\mu \neq 0$, that is, when the control system contains feedback. The slope of the control function is given by equation (112). Since $\bar{\psi}' = \chi' \mp \operatorname{sgn} F_1^*$ is a continuous function everywhere, the discontinuity of $F_1^{*\prime}$ is represented by the term $\mp \mu \operatorname{sgn} F_1^*$. Switch points *a* are defined as those where $\bar{\beta}'$ changes from negative to positive values; for these

$$F_1^{*\prime}(+0) - F_1^{*\prime}(-0) = +2\mu \tag{124a}$$

And switch points *b* are defined as those where $\bar{\beta}'$ changes from positive to negative values; for these

$$F_1^{*\prime}(+0) - F_1^{*\prime}(-0) = -2\mu \tag{124b}$$

Similarly as for position control (see Fig. 18) there are two types of

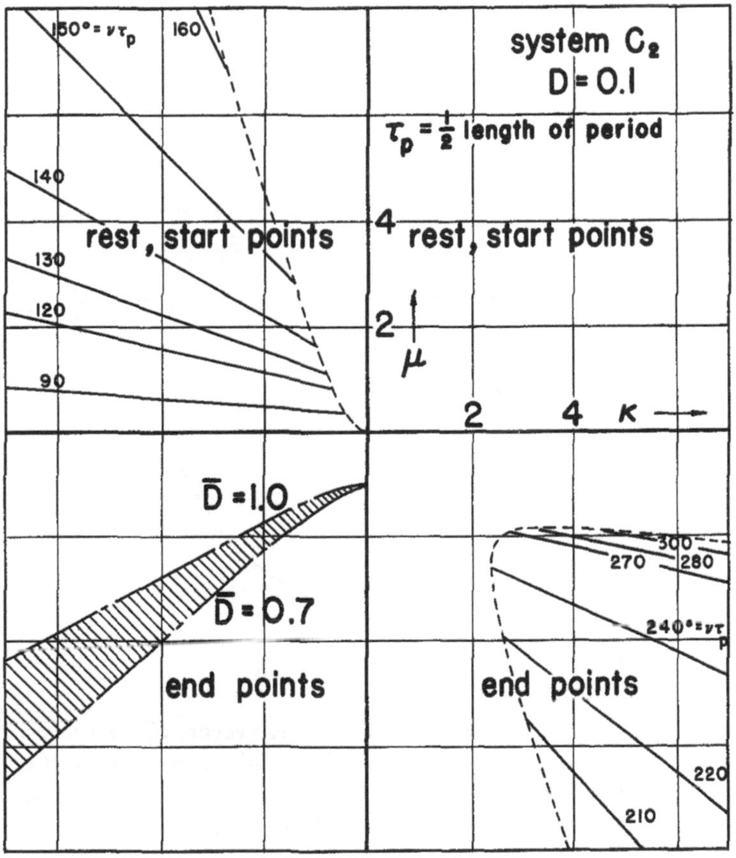

FIG. 51b

switching[1], governed by a simple refraction law. Therefore we may establish Table IV in order to have a complete survey. The break in the slope of F_1^* has the same consequences as in the theory of position control (see Fig. 18). Points occur in the switching plane $F_1^* = 0$ where the motion ceases to be regular (end points), and we also have points at which the motion can never arrive (starting points of a motion). It will prove worthwhile to note these regions in the κ, μ plane (Fig. 51), in which the possible periodic motions have already been described.

In order to prove that end points really occur (for example with system

[1] The type of switching is given by a or b and the system C_1 or C_2.

C_1 in the region $-\infty < \kappa < +\infty$, $\mu > 0$), one finds that the frames of Fig. 47 are very useful.

For the case $\mu = 0$ there is no discontinuity in the slope of F_1^*, and there-

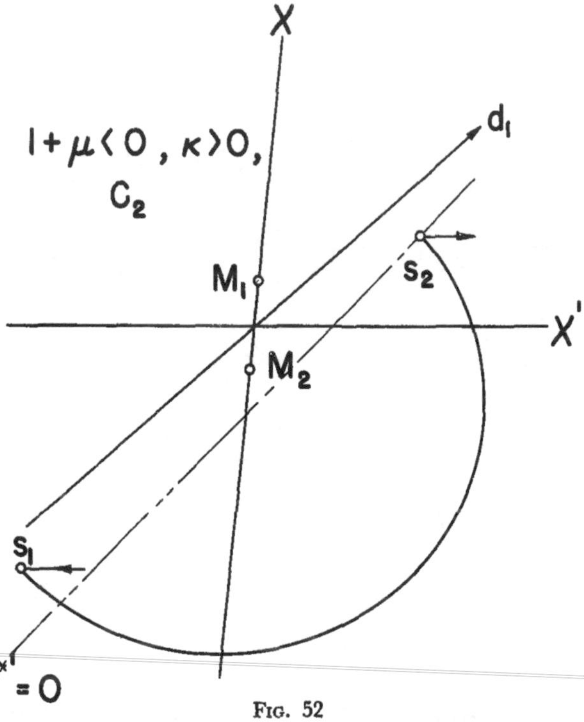

FIG. 52

fore no starting or end points can occur. However, F_1^* may be zero and have zero slope simultaneously. This point would denote the start of control at rest; Fig. 42 shows an example of this type. It is evident that control at rest may occur for motions with feedback (see Fig. 47f, for example).

Table IV

	System C_1	System C_2
$\mu > 0$	F^* breaks away from the perpendicular (end points possible)	F_1^* breaks towards the perpendicular (starting points possible)
$\mu < 0$	F_1^* breaks towards the perpendicular (starting points possible)	F_1^* breaks away from the perpendicular (end points possible)

The region (κ, μ) in which rest points occur may be found by looking at the projections of the phase curves in the $\bar{\beta}$, d_1 plane. Let us return to Figs.

47a-h. The projection of the phase curve in the d_1, $\bar{\beta}$ plane is a damped cosine line. In Fig. 47a a switchpoint A_s will *always* be followed by a second switchpoint B_s because the phase curve projection cannot fail to cut the

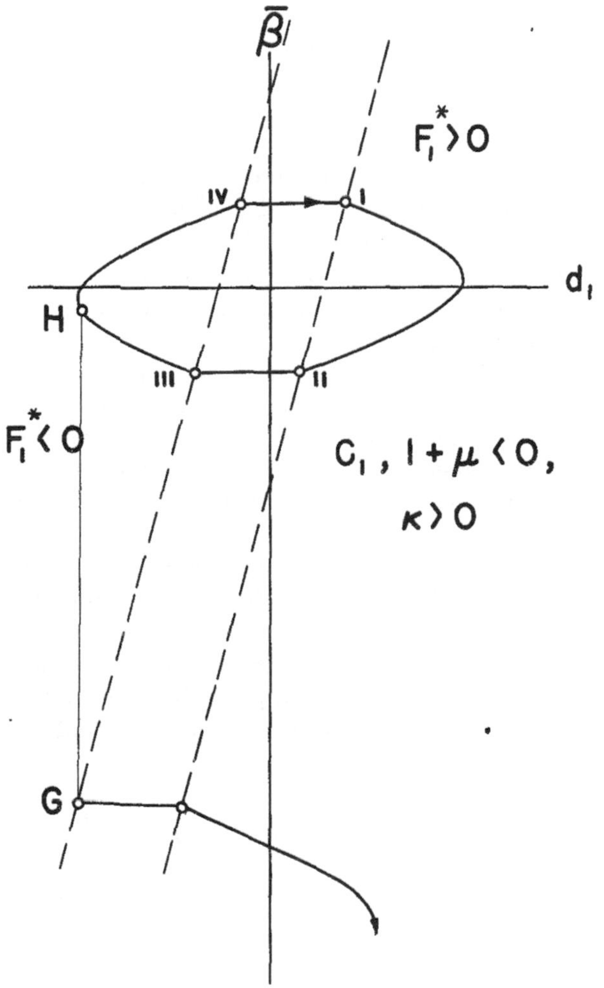

FIG. 53

switching line, and rest of control is impossible. This is not the case in Fig. 47b; here the phase curve projection leaving the point A_s might fail to cut the switching line if this line were slightly less steep than shown and the phase curve were the same. Therefore, a brief study of the construction frame for a particular example should enable one to predict whether or not rest of control will occur. The regions containing rest points thus found are noted in Figs. 51a, b.

The two conditions $F_1^* = 0$ and $F_1^{*\prime} = 0$ taken together define the limits of those regions where the special points (end, starting, and rest) occur. Since $F_1^{*\prime} = 0$ is given by equation (113b), all points having this quality will be in planes perpendicular to the χ, χ' plane. The intersection of these planes with the corresponding switching planes $F_1^* = 0$ result in two lines having these two conditions. The projection of these intersection lines into the d_1, $\bar{\beta}$ plane will coincide with the lines $F_1^* = 0$ in that plane. That is, each point in the lines $F_1^* = 0$ in the d_1, $\bar{\beta}$ plane may be the projection of a point with $F_1^{*\prime} = 0$.

Fig. 53 shows, for $(1 + \mu) < 0$, $\kappa > 0$ and system C_1, that a motion starting at a point G leads directly to rest of control. For the same combination of κ and μ, but a different value of $\bar{\beta}_i$, a periodic motion may be obtained. Note the phase curve I, II, III, IV, starting at the point H; the χ, χ' projection of H coincides with the χ, χ' projection of G. At once the enormous influence of the initial value of $\bar{\beta}$ is apparent. Without entering into detail, which might be interesting from the mathematical viewpoint but which is of little practical interest, it should be evident that the designer is urged to take $|\bar{\beta}_i| < |\bar{\beta}_{00}|$ if he is to avoid rest of control. This is always possible in a control mechanism: $\bar{\beta}_{00}$ is given by κ and μ. However, if D and ω of the basic system (Sec. 3.2) are given, then N and V are still to be determined. They may be chosen so that $|\bar{\beta}_i| < |\bar{\beta}_{00}|$ and χ and χ' do not exceed reasonable values.

Where a motion under ideal control would ordinarily come to an end point (see Figs. 43a, b), a motion under more realistic conditions (wherein time lags always exist) will not become undefined. Rather it will change to a high-frequency motion around an average path (Fig. 44a, b) which—assuming κ and μ have been chosen properly—will lead to $\bar{\psi} = 0$, $\bar{\psi}' = 0$, and $\bar{\beta} = 0$. If rapid damping is desired, it is essential to choose κ and μ so that the average motion

$$\bar{\psi}''_{\text{av}} + \bar{\psi}'_{\text{av}}\left(2D + \frac{\kappa}{\mu}\right) + \bar{\psi}_{\text{av}}\left(1 + \frac{1}{\mu}\right) = 0 \tag{108}$$

is a highly damped motion. The damping of this motion is characterized by

$$\bar{D} = \tfrac{1}{2}\,\frac{2D + \kappa/\mu}{\sqrt{1 + 1/\mu}} \tag{125}$$

If the oscillating motion changes to a motion of exponential character (aperiodic motion) then $\bar{D} \to 1$. For best results it is recommended that κ and μ be chosen such that

$$0.7 < \bar{D} < 1.0 \tag{126}$$

These regions are noted in Figs. 51a, b (shaded regions).

Thus Fig. 51 gives a survey of the types of motion which may occur in various κ, μ regions. There are regions in this diagram which indicate the possibility of more than one type of motion. Which type will actually occur depends upon the initial conditions of the motion (see Fig. 53).

In the theory of position control there were given diagrams (Figs. 23 and 26) which indicated in the plane of switch points ψ_s, ψ_s' the regions which could be end, starting, or rest points if κ and the system were chosen. Attempts have been made[1] to indicate those regions in the space $\bar{\psi}_s$, $\bar{\psi}_s'$, $\bar{\beta}_s$ for the velocity control, but the results are not nearly so satisfactory as for the case of position control.

5.5. Selection of the system and the coefficients κ and μ for a given mechanical problem. If the purpose of the control mechanism is to provide a rapid reduction of the disturbance to zero, it is advisable to choose coefficients κ and μ of the control mechanism such that for an ideal control end points would occur. Thus systems without feedback are excluded. Those regions where periodic motions as well as end points may occur should be avoided (see Figs. 51a, b). Therefore for system C_1 one should choose $\mu > 0$ and $\kappa > 0$; and for system C_2, $\mu < 0$ and $\kappa < 0$. But even these regions are somewhat restricted in that the selected values should provide a satisfactory damping coefficient, as we have just seen. The most desirable regions under these restrictions are those shown shaded in Figs. 51a, b. In these regions for κ and μ the discontinuous velocity control mechanism will work as a *continuous* mechanism if we consider only the average motion. The amplitude of the high-frequency motion which is superimposed is so small that it is of no importance. Its size depends solely on the time lags inherent in the mechanism (which should be kept small).[2]

In order that the end point region of the phase space be entered rapidly, the coefficients N and V of the mechanism have to be chosen properly. It is advisable to take them so that

$$|\bar{\beta}_i| < |\bar{\beta}_{00}| = \frac{|\kappa|}{|1 + \mu|}$$

and so that χ and χ' will not yield values of d_1 that are much larger than $2|d_{00}|$.

[1] See reference given in footnote 1 of this chapter on p. 53.
[2] Further details concerning the effects of time lags will be given in Chap. 7.

PART II

INFLUENCE OF CONTROL MECHANISM IMPERFECTIONS ON THE MOTION OF A BODY WITH A SINGLE DEGREE OF FREEDOM

In Chaps. 6 and 7 the influence of imperfections of the control mechanism upon its action and efficiency will be investigated.

In Sec. 2.3 the construction of control mechanisms and the sources of imperfections were discussed. Recalling those considerations, we now put forth the assumptions on which our studies will be based.

(1) The deviations ϕ and $\dot{\phi}$ are measured without error.

(2) The mixing device produces the control function

$$F = \rho_1\phi + \rho_2\dot{\phi} \text{ or } F = \rho_1\phi + \rho_2\dot{\phi} + \rho_3\beta$$

without error from the measured ϕ, $\dot{\phi}$, and the feedback β.

(3) The actual switching of the lever directing the control element position β, or its velocity V does not coincide with the zero points of F as it was assumed in the study of the ideal control system.

The primary objectives in the following analysis will be (1) to determine the location of the switch points in the phase plane or phase space, referred to the location of the zero points of F, and (2) to study the influence of this difference on the motion of the body.

6. EFFECTS UPON MOTION WITH POSITION CONTROL

The coincidence of the zero crossing of the control function F and the switching can be disturbed by three different types of imperfections. These are described in Sec. 2.3 and by Figs. 6-8, and they will be treated individually in the following sections.

6.1. Constant lag dependent upon the value of the control function (threshold). For this type of imperfection the control element changes its position when the control function $F = \rho_1\phi + \rho_2\dot{\phi}$ has reached a particular value σ_2 after passing zero (see Fig. 8).

With ideal control ϕ was replaced by ψ, and F by the function $F^* = \psi + \kappa\psi'$, because we were interested merely in the zero values of this function. Because (see equation (22))

$$F^* = \frac{1}{\rho_1}\frac{\omega^2}{N\beta_0}F \tag{127}$$

the threshold σ_2 of the function F must be replaced by

$$\sigma_2^* = \frac{1}{\rho_1}\frac{\omega^2}{N\beta_0}\sigma_2 \tag{128}$$

when using the control function F^*. For ideal control all results were given as a function of $\kappa = \omega\rho_2/\rho_1$, this being the essential parameter. With imperfect control we utilize the parameter σ_2/ρ_1 in addition to κ (as long as $\rho_1 \neq 0$). With increasing ρ_1 the value of σ_2^* decreases for constant σ_2, which means that imperfections will have less influence. On the other hand a small value of ρ_1 produces a large value of σ_2^*, which may become larger than unity.

The locus of the points $F^* = \psi_0 + \kappa\psi_0' = 0$ is a straight line in the phase plane ψ, ψ'. This line contains the point $(\psi, \psi') = (0, 0)$. The locus of the switch points

$$F^* = \psi_s + \kappa\psi_s' = \mp\sigma_2^* \tag{129}$$

are two lines parallel to $F^* = 0$ passing through the points $(\psi, \psi') = (\mp\sigma_2^*, 0)$. The distance between these lines is

$$2\Delta n = 2\sigma_2^* \frac{\sqrt{1 - D^2}}{\sqrt{1 - 2\kappa D + \kappa^2}} \tag{130}$$

or, since

$$\sigma_2^* = \frac{\sigma_2}{\rho_1} \frac{\omega^2}{N\beta_0} \text{ and } \kappa = \frac{\rho_2 \omega}{\rho_1}$$

there results

$$2\Delta n = \frac{2\sigma_2\sqrt{1 - D^2}}{\sqrt{\rho_1^2 - 2\rho_1\rho_2\omega D + \rho_2^2\omega^2}} \cdot \frac{\omega^2}{N\beta_0} \tag{131}$$

Equation (131) shows that $2\Delta n$ is finite, even for $\rho_1 = 0$ and $\rho_2 \neq 0$. The points $(\mp\sigma_2^*, 0)$ move toward infinity, but the lines defined by equation (129) are parallel to the ψ axis at a distance Δn from it.

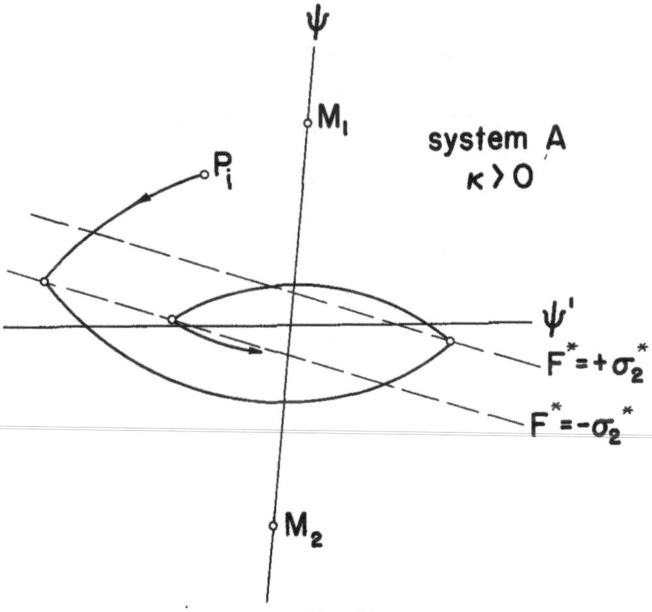

FIG. 54

A phase curve is shown in Fig. 54 for system A, $\kappa > 0$, and $\sigma_2^* = 0.15$. The two lines $F^* = \mp\sigma_2^*$ are used alternately according to the sign of $F^{*\prime}$ at the points $F^* = 0$ (see Fig. 8).

Since Δn generally will be small compared to unity it would appear that many phase curves will undergo only small changes. However there is one region in the phase plane where the delay in switching will have a very strong influence. That is the region where end points occur for ideal control. To understand this, let us look at the behavior of the control function F^* near a switch point for imperfect control. The discontinuity of the slope of F^* occurs at the actual switch points, at $F^* = \mp\sigma_2^*$. Hence the difficulty described in Sec. 4.3, referring to the upper left hand sketch of Fig. 18 for example, will never occur when this type of imperfection exists. Even if the entrance slope is in the shaded area and the law of refraction requires an exit slope which is positive, no obstacle to continuing the motion exists, for end points do *not* exist. This is clearly shown in Fig. 55. If $F^{*\prime}(+0) \geqslant 0$ as in Fig. 55, it may happen that the motion never leaves the zone $|F^*| \leqslant \sigma_2^*$ (see Fig. 56) in passing over to a periodic motion.[1] Since the points where $F^* = 0$ and where switching occurs do not coincide, points called "starting points" no longer exist.

The periodic motions of a controlled system, having the type of imperfection with which we are now concerned, will be given by equations (82, 83a, b). However, the value of κ corresponding to a given $\nu\tau_p$ is determined by the relation

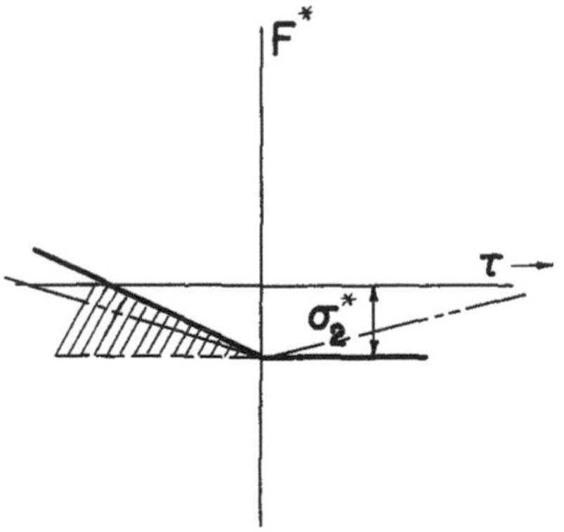

FIG. 55

$$\frac{1}{\mp \operatorname{sgn} F^*} (\psi_p + \kappa\psi_p') = -\sigma_2^* \tag{132a}$$

for a switch point *a* in system A for example, or

$$\frac{D}{\sqrt{1-D^2}} \sin \nu\tau_p - \sinh D\tau_p - \frac{\kappa}{\sqrt{1-D^2}} \sin \nu\tau_p$$
$$= -\sigma_2^* (\cos \nu\tau_p + \cosh D\tau_p) \tag{132b}$$

In those regions where the final periodic state depends solely upon the imperfection and its existence, the length of period strongly depends upon σ_2^*. If, in system A, κ is not very small, a small value of σ_2^* will result in a small value of τ_p. As a first approximation we may write

$$\tau_p = \frac{+2\sigma_2^*}{\kappa} = \frac{2\sigma_2(\omega^2/N\beta_0)}{\rho_2\omega}, \quad \kappa \geqslant 1, \quad D \geqslant 0, \quad \sigma_2^* << 1 \tag{133}$$

The second form shows that τ_p is finite for $\rho_1 \to 0$ $(\kappa \to \infty)$.

---

[1] The switch points of the periodic motion do not lie exactly on the axis ψ'. However, the deviation is so small that it is impossible to show the difference in a drawing. This is also the reason why it looks as if the periodic motion is attained after one cycle; in reality it will take many cycles to come close to the asymptotic state.

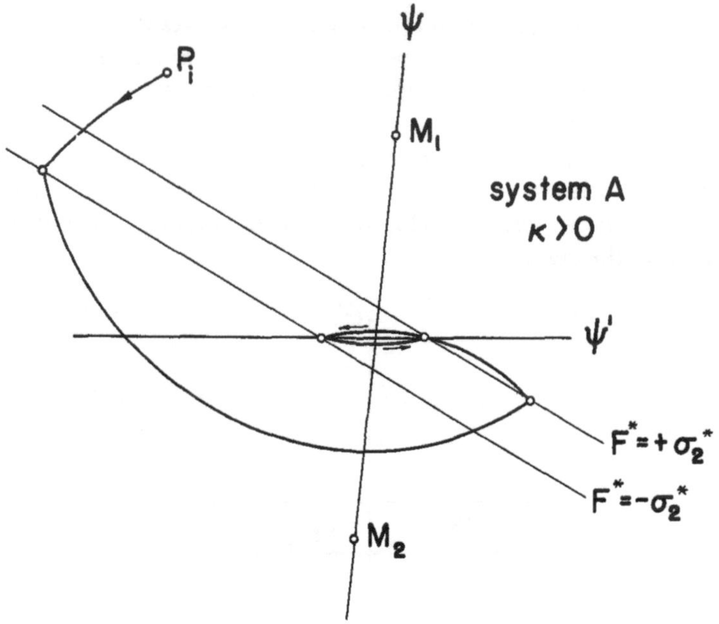

FIG. 56

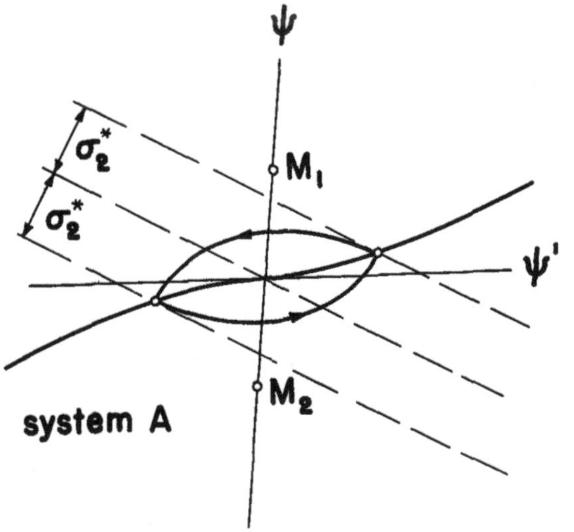

FIG. 57

If σ_2^* is larger, and the periodic motion is not necessarily the result of the imperfection, then a different method of solution of equation (132a) is required. The easiest way to solve this equation is to draw in the ψ, ψ' plane for a fixed D the curve of the periodic switch points for varying κ (use Fig. 29 for this purpose). Then trace the line $F^* = \sigma_2^*$ (see Fig. 57, system A). The intersection gives the periodic solution for a given κ, D,

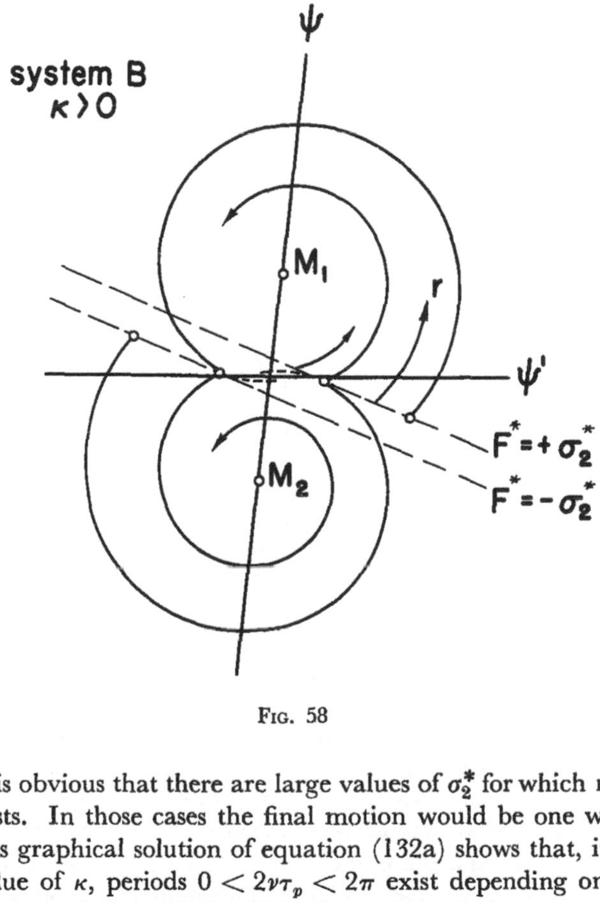

FIG. 58

and σ_2^*. It is obvious that there are large values of σ_2^* for which no periodic solution exists. In those cases the final motion would be one with control at rest. This graphical solution of equation (132a) shows that, in system A for each value of κ, periods $0 < 2\nu\tau_p < 2\pi$ exist depending on the value of σ_2^*. For system B the periods will be larger than $2\pi/\nu$.

The preceding analysis shows that imperfections of the type considered here will have a strong but favorable influence only in those regions where the perfect control system would lead to end points in system A. A general survey of the types of motion which may occur is easily given (following the pattern established in Sec. 4.6) by applying modifications to the ideal control system for an imperfection $\sigma_2^* << 1$.

System A: positive κ $(0 < \eta < \sigma)$. For $\sigma_2^* << 1$ every motion will change to a high-frequency periodic one after entering a region where end points would normally occur in an ideal control system. The period is given by equation (133) and the considerations following. (If σ_2^* is very large, rest of control may occur; thus a large σ_2^* is to be avoided.)

System A: negative κ ($\sigma < \eta < \pi$). Again for $\sigma_2^* << 1$ every motion will tend toward a periodic one. The length of period $2\tau_p$ will be less than $2\pi/\nu$; in the limit it equals $2\pi/\nu$. The length of period $2\tau_p$ is given by equation (132b) as a function of κ and σ_2^*. (Again for very large σ_2^* rest of control may occur.)

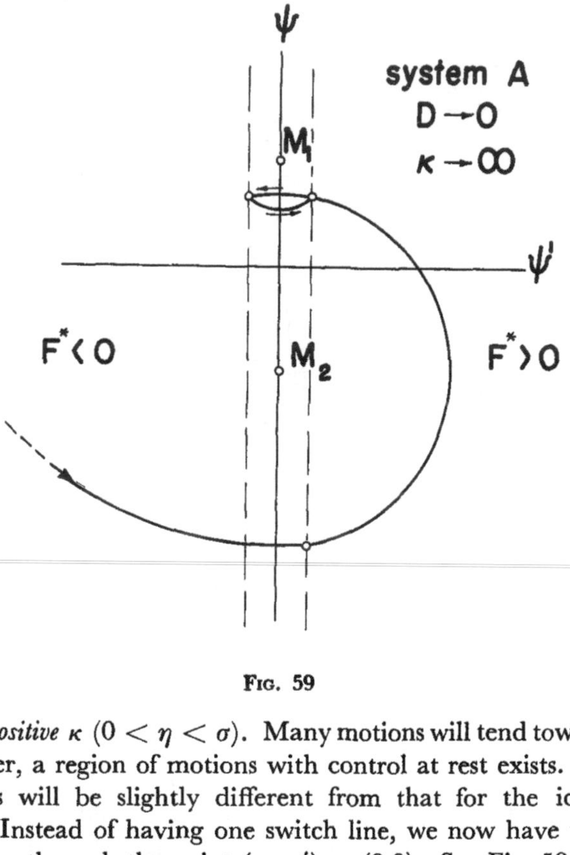

FIG. 59

System B: positive κ ($0 < \eta < \sigma$). Many motions will tend toward periodic ones; however, a region of motions with control at rest exists. The region of rest points will be slightly different from that for the ideal control mechanism. Instead of having one switch line, we now have two, neither of which passes through the point $(\psi, \psi') = (0,0)$. See Fig. 58.

System B: negative κ ($\sigma < \eta < \pi$). The motion of the body will tend toward one with control at rest—including perhaps a motion of high frequency for brief intervals of time (see Fig. 33).

The regions of rest points for the particular case of $D = 0$ have been studied rather completely by K. Scholz.[1]

An interesting example of a periodic motion occurs when $D \to 0$ and $\kappa \to \infty$ (see Fig. 59). This periodic motion consists of two parts, which are *not*

[1] Über Bewegungen eines Schwingers unter dem Einfluss von Schwarz-Weiss-Steuerungen. II. Bewegungen eines Schwingers mit Stellungszuordnung mit Schaltverschiebungen. *Zentrale für wissenschaftliches Berichtswessen der Luftfahrtforschung des Generalluftzeugmeisters* (*ZWB*), Untersuchungen und Mitteilungen Nr. 1327, Berlin, January 2, 1945.

traversed in equal intervals of time as has been the case in all previous examples.

6.2. Direct time lag. This type of imperfection is characterized by the fact that the control element changes its position a constant time t_r after F passes zero (see Fig. 7).[1]

Let ψ_0 and ψ_0' correspond to $F^* = 0$ so that

$$\psi_0 + \kappa\psi_0' = 0 \tag{134}$$

At some time $\tau = \tau_r$ later ψ and ψ' will have the values ψ_s and ψ_s' (values

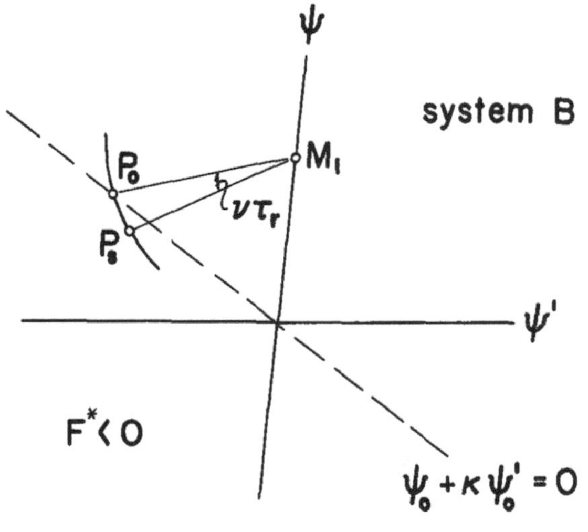

FIG. 60

at the switch points). The points $\psi_0 + \kappa\psi_0' = 0$ lie on a straight line through the origin of the coordinate system (ψ, ψ'). The locus of the points ψ_s and ψ_s' is as yet undetermined. Immediately we see that a certain threshold σ_2^* corresponds to a certain time lag τ_r

$$\sigma_2^* \cong \tau_r \left|\frac{dF}{d\tau}\right|_{F^*=0} \tag{135}$$

which suggests an analogy between the influences of the two different types of imperfections.

Assuming that $\tau = 0$ at ψ_0, ψ_0' and $\tau = \tau_r$ at ψ_s, ψ_s' we may express the coordinates of two corresponding points P_0 and P_s as follows (see Fig. 60 and equations (50, 51)):

$$\left.\begin{aligned} \psi_s &= 2C\, e^{-D\tau_r} \cos(\nu\tau_r + \varepsilon) \mp \operatorname{sgn}(-F^*) \\ \psi_s' &= 2C\, e^{-D\tau_r} \cos(\nu\tau_r + \varepsilon + \sigma) \end{aligned}\right\} \tag{136}$$

[1] The differential equation governing this problem is (compare with equation (27))

$$\psi''(\tau) + 2D\psi'(\tau) + \psi(\tau) = \mp \operatorname{sgn}[\psi(\tau - \tau_r) + \kappa\psi'(\tau - \tau_r)]$$

and is known as a difference-differential equation. This type of equation is often treated analytically. However, we shall see that its solution may be obtained easily by use of phase curves.

and

$$\left.\begin{aligned}\psi_0 &= 2C\cos\varepsilon \mp \operatorname{sgn}(-F^*)\\ \psi_0' &= 2C\cos(\varepsilon+\sigma)\end{aligned}\right\}\tag{137}$$

Use equations (136) to eliminate the constants C and ε from equations (137); now we obtain ψ_0 and ψ_0' as functions of ψ_s, ψ_s', and τ_r. These new relations may then be introduced into equation (134). After a simple but tedious transformation equation (138) is obtained:

$$\psi_s \sin[\sigma - (\eta - \nu\tau_r)] + \psi_s' \sin(\eta - \nu\tau_r)$$
$$\mp \operatorname{sgn}(-F^*)\{e^{-D\tau_r}\sin(\sigma-\eta) - \sin[\sigma - (\eta - \nu\tau_r)]\} = 0 \tag{138}$$

where

$$\eta = \operatorname{arctg}\frac{\kappa\sqrt{1-D^2}}{1-\kappa D}\tag{54}$$

Now we recognize that the switch points ψ_s, ψ_s' lie on two parallel lines which form an angle $(\eta - \nu\tau_r)$ with the negative ψ' axis and an angle $\nu\tau_r$ with the straight line $\psi_0 + \kappa\psi_0' = 0$. The distance between these two switch lines is

$$2\Delta n = -2\{e^{-D\tau_r}\sin(\sigma-\eta) - \sin[\sigma-(\eta-\nu\tau_r)]\}\tag{139}$$

For small values of τ_r: $2\Delta n = 2\tau_r \sin\eta$. This analysis allows us to establish the following analogy between the two types of imperfections:

Table V

Type of imperfection	Locus of switch-points	Distance between switch lines	Angle between switch lines and negative ψ' axis
threshold (i.e. σ_2^*)	Two parallel straight lines	$\dfrac{2\sigma_2\sqrt{1-D^2}}{\sqrt{\rho_1^2 - 2\rho_1\rho_2\omega D + \rho_2^2\omega^2}}\cdot\dfrac{\omega^2}{N\beta_0}$	η
Time lag τ_r	Two parallel straight lines	Equation (139) approx.: $2\tau_r \sin\eta$ $= 2\omega t_r \sin\eta$	$\eta - \nu\tau_r$

The straight lines corresponding to different values of η having fixed τ_r pass through a fixed point having the coordinates

$$\left.\begin{aligned}\psi^* &= \mp\operatorname{sgn}(-F^*)\left[1 - e^{-D\nu\tau_r}\frac{\sin(\sigma-\nu\tau_r)}{\sin\sigma}\right]\\ \psi^{*\prime} &= \mp\operatorname{sgn}(-F^*)\,e^{-D\nu\tau_r}\frac{\sin\nu\tau_r}{\sin\sigma}\end{aligned}\right\}\tag{140}$$

If $\tau_r \to 0$, the coordinates of this fixed point ψ^*, $\psi^* \to 0, 0$ because the switch lines tend towards the line $\psi_0 + \kappa\psi_0' = 0$.

With all of the foregoing analysis at hand let us study some examples of phase curves. Figs. 61 and 62 give examples for system A with $\kappa > 0$ and $\kappa < 0$, respectively. Fig. 63 shows a special example for system A in which κ is positive but the angle $(\eta - \nu\tau_r)$ is negative. In this last example note

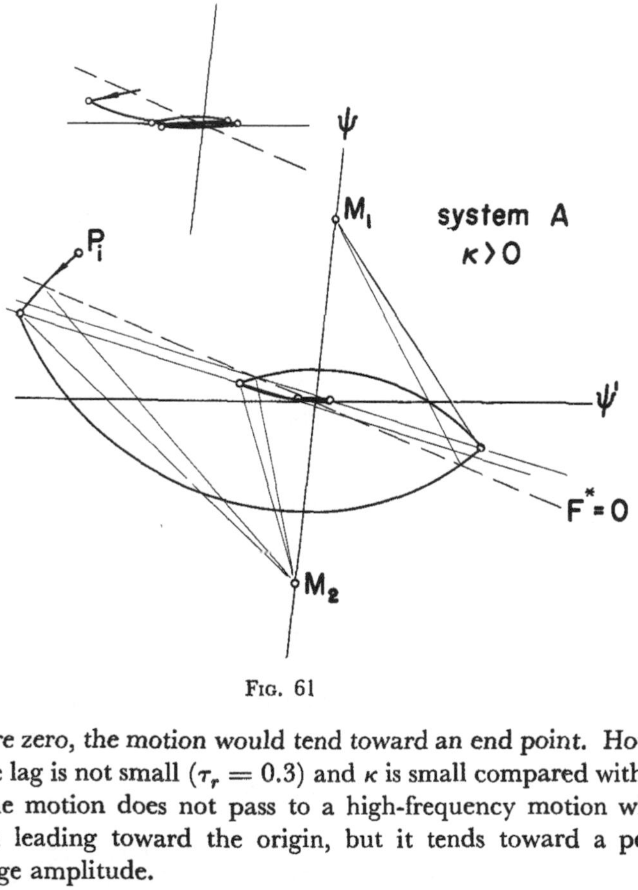

FIG. 61

that if $\nu\tau_r$ were zero, the motion would tend toward an end point. However, since the time lag is not small ($\tau_r = 0.3$) and κ is small compared with unity ($\kappa = 0.2$), the motion does not pass to a high-frequency motion with an average path leading toward the origin, but it tends toward a periodic motion of large amplitude.

The periodic motions are given by equation (83a, b). The value of κ (or η) corresponding to a certain τ_p is given by (corresponding to equation (138))

$$\psi_p \sin [\sigma - (\eta - \nu\tau_r)] + \psi_p' \sin (\eta - \nu\tau_r)$$
$$\mp \operatorname{sgn} (-F^*) \{e^{-D\tau_r} \sin (\sigma - \eta) - \sin [\sigma - (\eta - \nu\tau_r)]\} = 0 \qquad (141)$$

Or, after introducing ψ_p and ψ_p' and performing a simple transformation, we obtain

$$\sin \nu\tau_p \cos (\sigma - \eta + \nu\tau_r) + \sinh D\tau_p \sin (\sigma - \eta + \nu\tau_r)$$
$$- (\cos \nu\tau_p + \cosh D\tau_p) [e^{-D\tau_r} \sin (\sigma - \eta) - \sin (\sigma - \eta + \nu\tau_r)] = 0 \qquad (142)$$

As in the case of threshold imperfection, end points cannot occur in mechanisms having time lags because there will always be a possibility of continuing the motion. The regions where end points would occur in an ideal mechanism will be regions of considerable importance. Looking at Fig. 61, we see that the final state of the motion in these regions (if

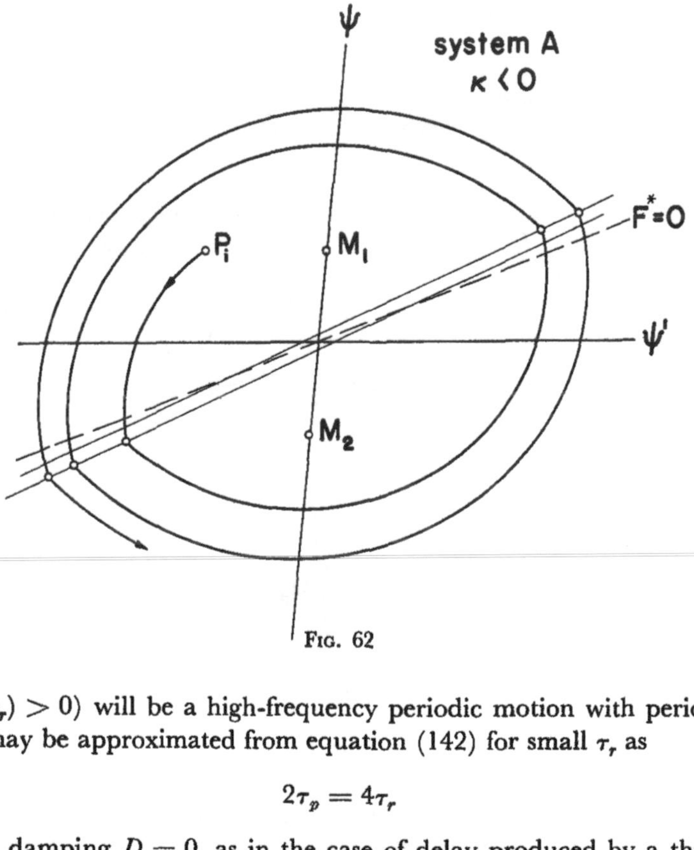

FIG. 62

$(\eta - \nu\tau_r) > 0$) will be a high-frequency periodic motion with period $2\tau_p$, which may be approximated from equation (142) for small τ_r as

$$2\tau_p = 4\tau_r \tag{143}$$

If the damping $D = 0$, as in the case of delay produced by a threshold imperfection, a periodic motion exists which does not consist of two symmetric parts. It occurs for $\eta - \nu\tau_r = \pi/2$, which means that $\kappa = -\cot \nu\tau_r$.

A body controlled by a mechanism having constant time lags will undergo the same types of motion as a body controlled by a mechanism having threshold imperfection. Therefore, if in the survey at the end of Sec. 6.1 we replace η by $\eta - \nu\tau_r$ and change κ correspondingly, we shall have a complete survey of the motion of a system having time lag. We must also note that equations (132a, 133) are to be replaced by equations (141, 143), respectively.

Up to this point we have studied only the influence of a real time lag upon the system. The idea might occur that there exist some possibilities for the instrument to compensate or even to overcompensate for such a delay. This would result in having an advanced influence instead of a delay which means a negative retardation time. By looking at a figure corresponding to Fig. 55, one can easily see that this is not an advantage. All of the disadvantages of the ideal system remain.

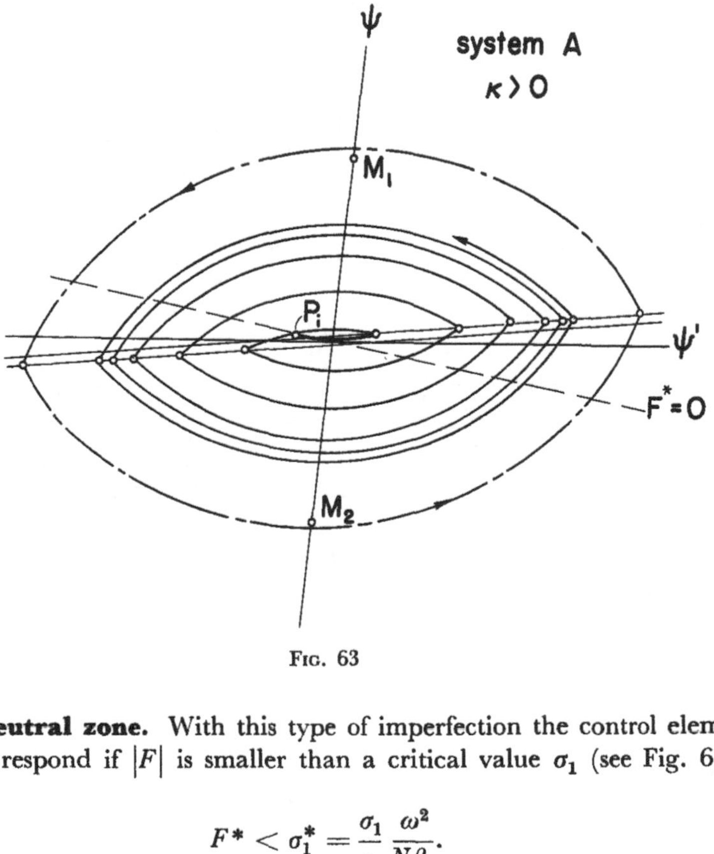

FIG. 63

6.3. Neutral zone. With this type of imperfection the control element does not respond if $|F|$ is smaller than a critical value σ_1 (see Fig. 6) or

$$F^* < \sigma_1^* = \frac{\sigma_1}{p_1} \frac{\omega^2}{N\beta_0}.$$

The meaning of this statement may be explained by an example. Choose system A with $\kappa > 0$ (Fig. 64). Let the motion begin at the point P_i where $F^* > 0$ and follow a spiral path with center at M_2. At the point a_1 the control function reaches the value $+\sigma_1^*$, and the control element switches to the neutral position. Between the points a_1 and a_2 the phase curve is a segment of a spiral about the origin. At the point a_2 the control function reaches the value $-\sigma_1^*$; immediately the control element resumes its action. The path of the motion is then a spiral about M_1. At the point b_1 where $F^* = -\sigma_1^*$ the body starts moving about the origin again, and so on. If $\sigma_1^* \to 0$, the points a_1 and a_2 approach each other. If $\sigma_1^* = 0$, these two points coincide. For $\sigma_1^* = 0$ (ideal control), the example shown in Fig. 64 would appear the same as that of Fig. 13 (S_1 would be of type a and S_2 of type b).

The points a_1, a_2, b_1, and b_2 in Fig. 64 are regular switch points. However, the next point a_1' is the last switch point after which motion with control at rest ensues. Looking at the motion as a whole we see that the motion is nevertheless more damped than the motion with no control at all.

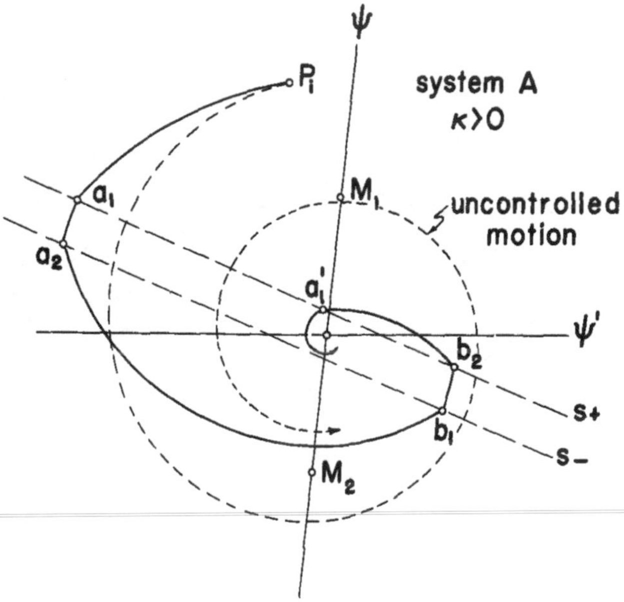

FIG. 64

For this case of imperfection the switch points will lie on two lines parallel to $F^* = 0$, each at a distance Δn from this line.

$$\Delta n = \sigma_1^* \frac{\sqrt{1 - D^2}}{\sqrt{1 - 2\kappa D + \kappa^2}} = \sigma_1 \frac{\sqrt{1 - D^2}}{\sqrt{1 - 2\kappa D + \kappa^2}} \left[\frac{1}{\rho_1} \frac{\omega^2}{N\beta_0}\right] \qquad (144)$$

$$= \frac{\sigma_1 \sqrt{1 - D^2}}{\sqrt{\rho_1^2 - 2\rho_1\rho_2\omega D + \rho_2^2\omega^2}} \frac{\omega^2}{N\beta_0}$$

As explained on pp. 19 and 52, the factor $N\beta_0/\omega^2$ determines the unit of ψ. Thus for a given D the three essential parameters of the problem are ρ_1, $\omega\rho_2$, and σ_1.

The following table shows the values of

$$\Delta n' = \sigma_1 \frac{\sqrt{1 - D^2}}{\sqrt{\rho_1^2 - 2\rho_1\rho_2\omega D + \rho_2^2\omega^2}}$$

for special values of these parameters:

Table VI

σ_1	ρ_1	$\rho_2\omega$	$\Delta n'$
0	ρ_1	$\rho_2\omega$	0
σ_1	0	$\rho_2\omega$	$\sigma_1 \frac{\sqrt{1 - D^2}}{\rho_2\omega}$
σ_1	ρ_1	0	$\sigma_1 \frac{\sqrt{1 - D^2}}{\rho_1}$

The two combinations

(1) $\rho_1 = 1, \quad 0 \leqslant |\omega\rho_2| \leqslant 1$

(2) $\omega\rho_2 = 1, \quad 0 \leqslant |\rho_1| \leqslant 1$

describe all possible variations for a linear control function F. Thus we obtain

$$\sigma_1 \sqrt{\frac{1 - D}{2}} \leqslant \Delta n' \leqslant \sigma_1$$

The influence of a neutral zone will be studied by keeping

$$\Delta n' = \frac{\sigma_1 \sqrt{1-D^2}}{\sqrt{\rho_1^2 - 2\rho_1\rho_2\omega D + \rho_2^2\omega^2}} = \text{const.} \tag{145}$$

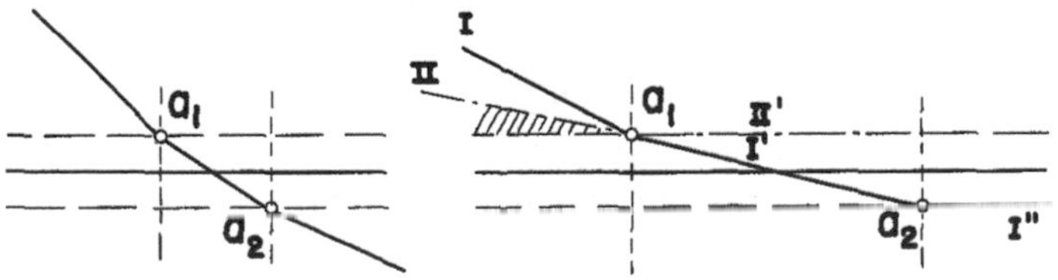

FIG. 64a

The switch line adjacent to $F^* > 0$ will be called S_+, and that adjacent to $F^* < 0$ will be called S_-.

An example for system A with $\kappa > 0$ is given in Fig. 64.

In all switch points the control function has a discontinuous slope corresponding to the action of the control element, as shown in Fig. 6. Therefore in switch points

$$\left.\begin{array}{lll} \text{of type} & a_1, a_2: & F^{*\prime}(+0) - F^{*\prime}(-0) = \kappa \\ \text{and of type} & b_1, b_2: & F^{*\prime}(-0) - F^{*\prime}(-0) = -\kappa \end{array}\right\} \quad (146)$$

This discontinuity in the slope of F^* leads to the existence of end and starting points on the switch lines, as is clearly shown in Fig. 64a. Comparison of Fig. 64a with Fig. 55 shows the essential difference between the two types

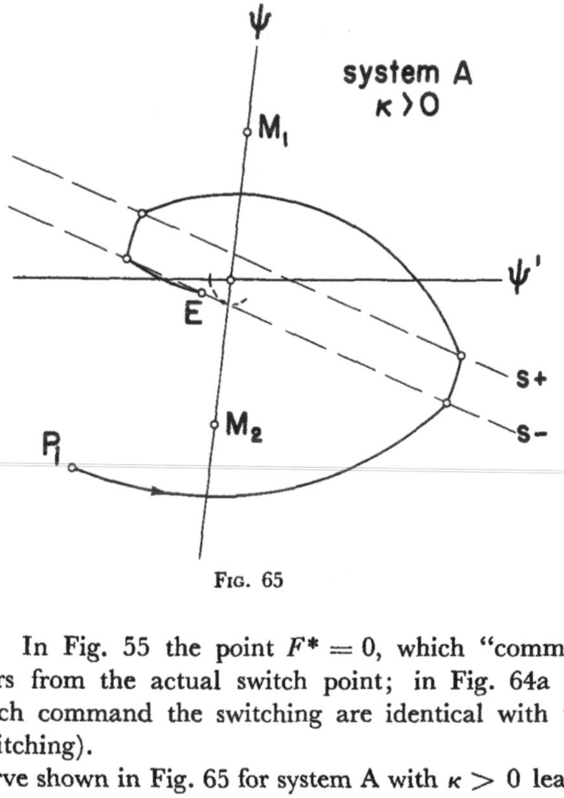

FIG. 65

of imperfection. In Fig. 55 the point $F^* = 0$, which "commands" the switching, differs from the actual switch point; in Fig. 64a the points $F^* = \pm\sigma_1^*$ which command the switching are identical with the actual change in β (switching).

The phase curve shown in Fig. 65 for system A with $\kappa > 0$ leads toward an end point. On the other hand the phase curve shown in Fig. 66 for system A with $\kappa < 0$ originates at a starting point.

Regions for end points and starting points (type a_2) for varying κ of system A are shown in Figs. 68a, b with $\Delta n'$ held constant. The reason for holding this quantity constant instead of σ_1^* is apparent from the immediately preceding discussion of the influence of ρ_1 and $\rho_2\omega$. In case $\Delta n' \to 0$ the regions shown in these figures (when combined into one figure) would appear as those in the upper half of Fig. 23a. Also shown in Figs. 68a, b are regions of rest points (type a_1) as they appear in Fig. 64. For system A such points may never occur in an ideal control mechanism.

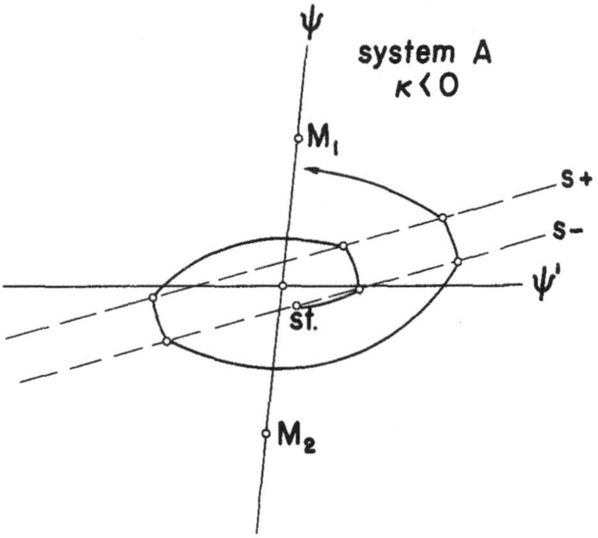

FIG. 66

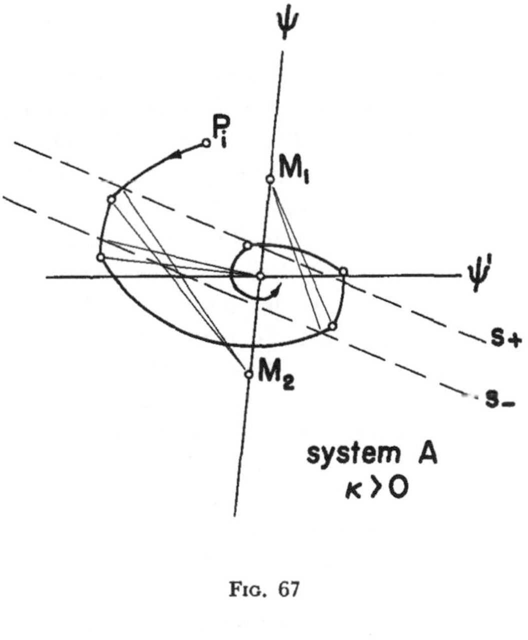

FIG. 67

An example of a motion for system B with $\kappa < 0$ is given in Fig. 69. Depending upon various initial points, the motion may tend either to end points or to rest points. It is interesting to note that control at rest may occur about any one of three different centers: M_1, M_2, or the origin of the coordinate system.

As seen from Fig. 70a rest points and end points may arise for motion in system B ($\kappa < 0$). Switch points of type b_2 located along the S_- line between the points S_{e_2} and S_{r_2} will lead to rest of control about M_2. Switch points

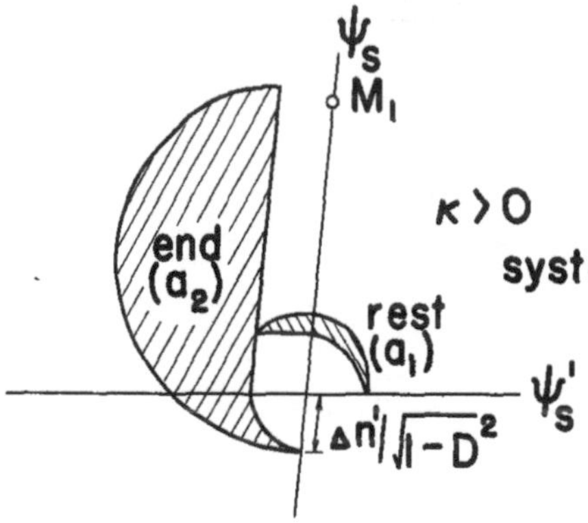

FIG. 68a

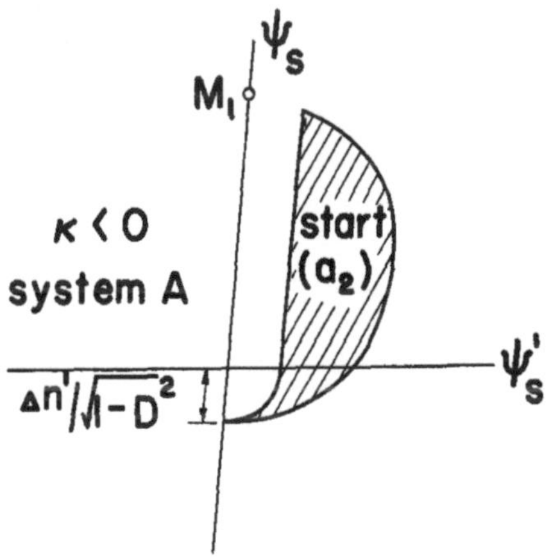

FIG. 68b

of type a_1 located along the S_- line between the points S_{e_2} and O' will be end points of the motion. From Fig. 69 it is interesting to note that end points of type a_1 may be end points of type b_2 for other phase curves. Fig. 70*b* shows the regions of end points and rest points (type b_2) for varying negative κ ($\Delta n' =$ constant). An additional region of rest points also exists: it is composed of type b_1 switch points leading to rest of control about the origin in a manner very similar to that already studied in system A.

Fig. 71 is analogous to Fig. 69 for positive κ. We see that both starting points and rest points may exist. Further, there may even be points which are both starting and rest points. Again, Figs. 72a, b are analogous to Figs. 70a, b in that they show the various regions of motion and their limits. The region of starting points of type b_1 which corresponds to the rest points of type b_1 in Figs. 70a, b is not shown in Fig. 72b.

In Figs. 68a, b, 70b, and 72b, only regions for certain types of end points, starting, and rest points are marked. The analogous regions for end points b_2 in Fig. 68a for instance may be found either by symmetry considerations or by detailed construction.

If the neutral zone type of imperfection is coupled with one of the other types characterized by delayed response, then the unfavorable influence of the discontinuities disappears and no points exist where the motion of the body becomes undefined (end points). An example of this type is given in Fig. 67. Here the motion, controlled by a mechanism having both neutral zone and time lag imperfection, leads to a spiral around $\psi = 0$, $\psi' = 0$ (compare Fig. 65).

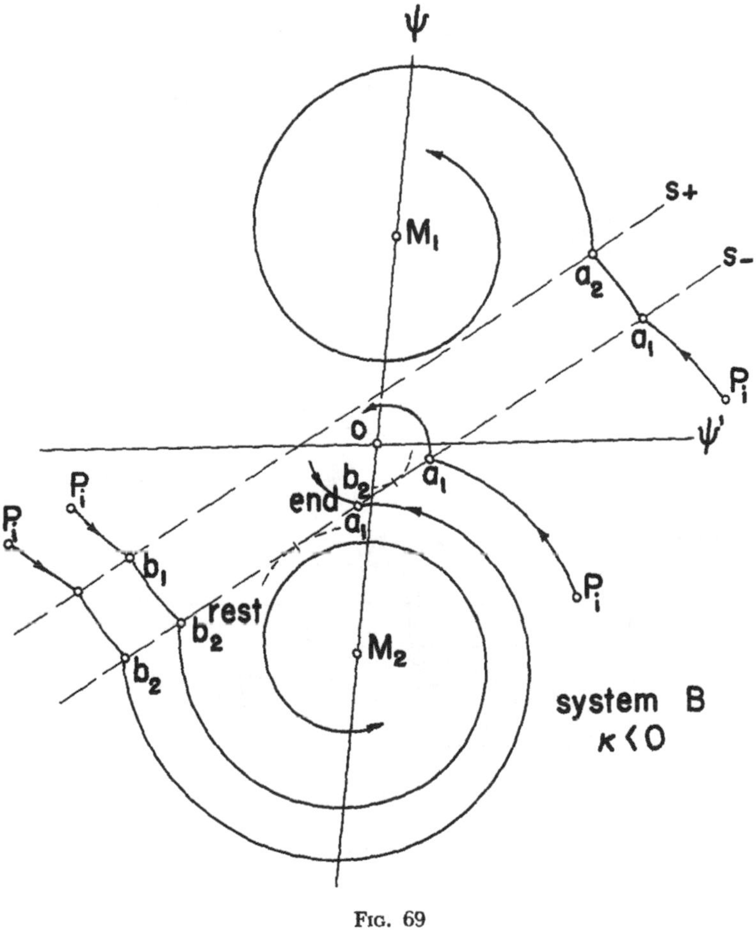

FIG. 69

As with ideal control systems, a periodic motion may exist with "neutral zone" imperfections. Such a periodic motion is shown in Fig. 73 for system A with $\kappa < 0$. Any motion controlled by a system A mechanism and having these particular values of κ and σ^* and initial ψ and ψ' values outside of the shaded area of the neutral zone will always tend towards this periodic motion.

The computation of the periodic motion, which may occur in system A for negative κ and in system B for positive κ, is complicated. Consider a

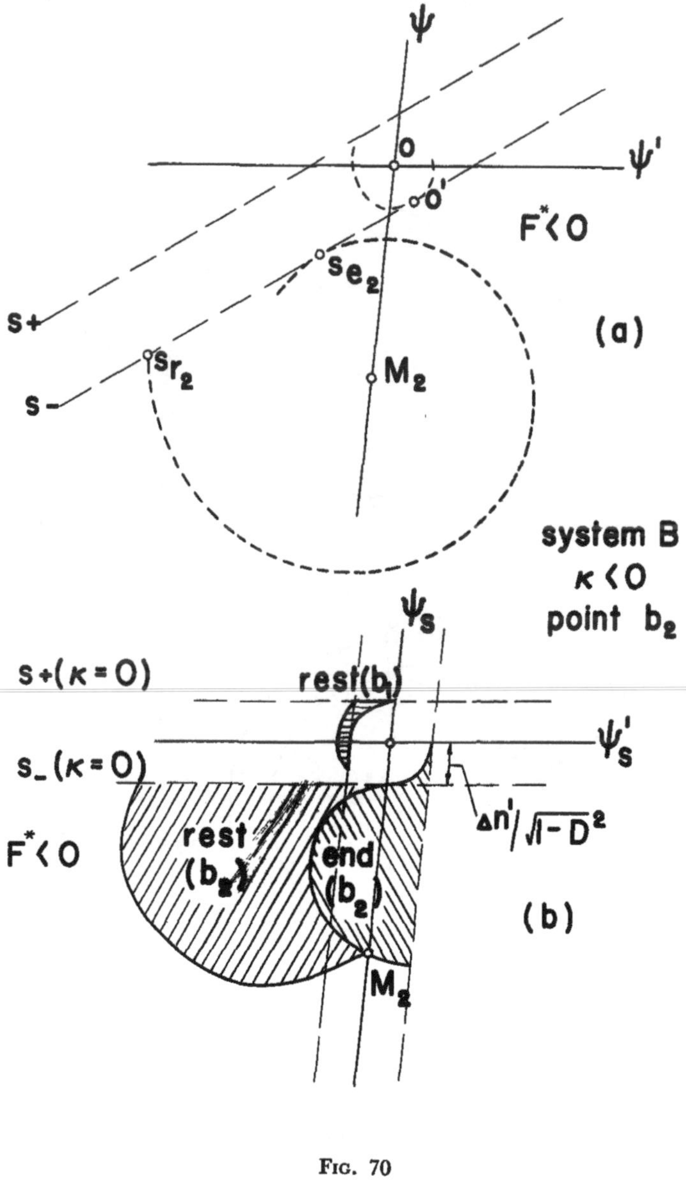
ψ
O
ψ'
O'
F* < O
s_{e_2}
S+
(a)
s_{r_2}
M_2
S −
system B
κ < O
point b_2
ψ_s
S+(κ = O)
rest(b_1)
ψ_s'
S_(κ = O)
$\Delta n'/\sqrt{1-D^2}$
rest
(b_2)
F* < O
end
(b_2)
(b)
M_2

FIG. 70

periodic motion (using the notation of Fig. 74) which is composed of segments of spirals as follows:

$\overline{S_{11}S_{12}}$ is given by a spiral around the origin O

$\overline{S_{21}S_{22}}$ is given by a spiral around the origin O

$\overline{S_{12}S_{21}}$ is given by a spiral around M_2 (system A)

$\overline{S_{22}S_{11}}$ is given by a spiral around M_1 (system A)

Also note that $\overline{OS}_{12} = \overline{OS}_{22}$; $\overline{OS}_{21} = \overline{OS}_{11}$.

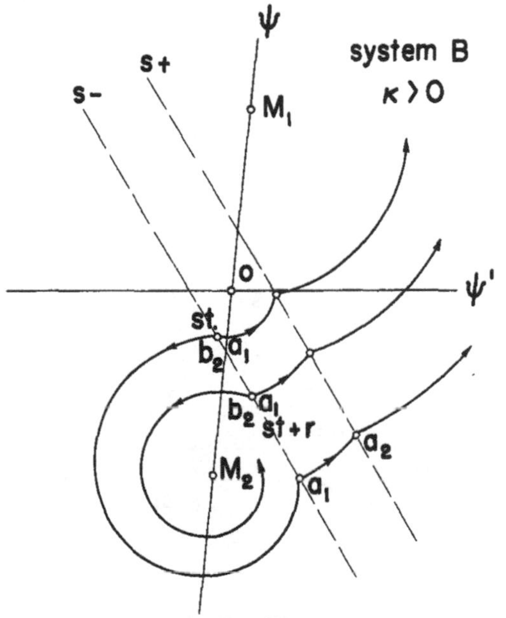

FIG. 71

In dealing with ideal control mechanisms we found that the simplest procedure was to assume a period and then compute κ or η. Here the procedure is considerably more complicated, but the results of such a computation indicate that when σ_1^* is small, the periodic motion obtained is close to that of the ideal control. But there is one essential difference between the two periodic motions: there are no periodic motions having very small periods for finite values of σ_1^* because phase curves near the origin have only rest of control (see Fig. 64). For instance, any motion starting at a point (ψ_i, ψ_i') with (see Fig. 73)

$$\frac{\Delta n'}{\sqrt{1 - D^2}} \geqslant \sqrt{\psi_i^2 + \psi_i'^2 + 2D\psi_i\psi_i'}$$

or in the shaded region of Fig. 68a, will be uncontrolled. In this case the chosen value of κ naturally has no influence. However, with ideal control the period tends towards zero with κ (see Fig. 29).

The above relation may be useful in determining experimentally the size of the neutral zone ($2\sigma_1$) (see equation (145)). When $\Delta n'/\sqrt{1-D^2}$ is

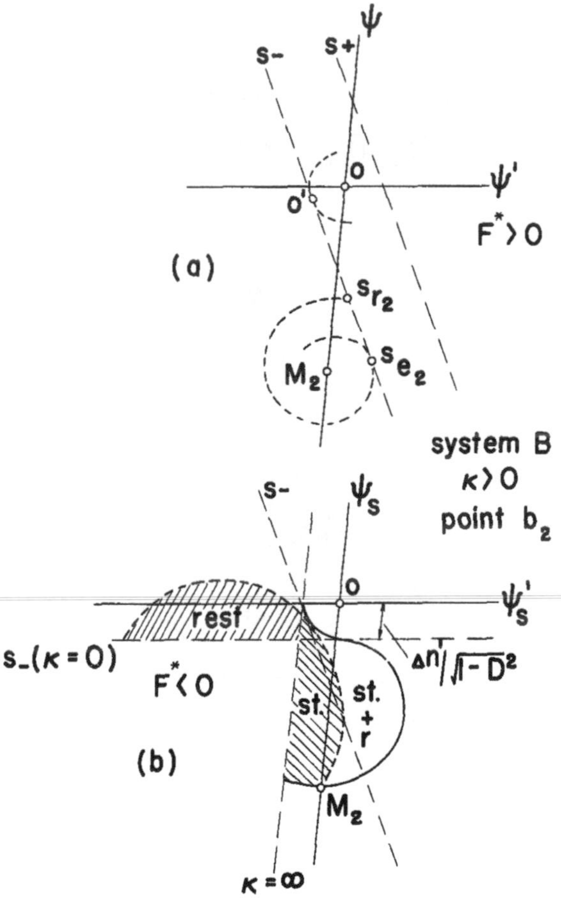

FIG. 72

larger than the radical $\sqrt{\psi_i^2 + \psi_i'^2 + 2D\psi_i\psi_i'}$, the control mechanism will have no influence upon the motion of the body.

6.4. Relative importance of the three types of imperfections. This chapter has dealt with three types of control mechanism imperfections and their influences upon the motion of a controlled body. Two of these types of imperfection (described in Sec. 6.1 and 6.2) have been found to possess a favorable influence in the reduction of a disturbance of the regular motion as long as these imperfections are small. As already indicated (see Sec. 4.7), system A with $\kappa > 0$ should be selected if a quick damping of

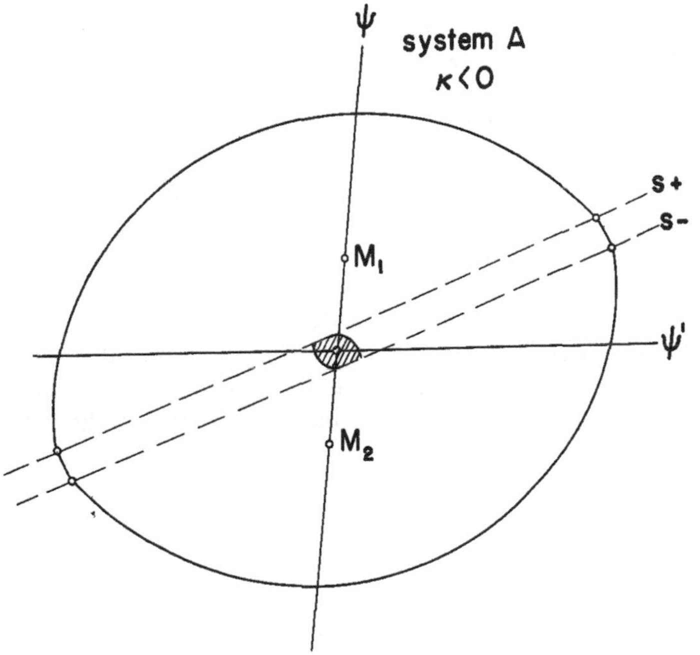
ψ
system A
κ<0
s+
s-
M₁
ψ'
M₂

FIG. 73

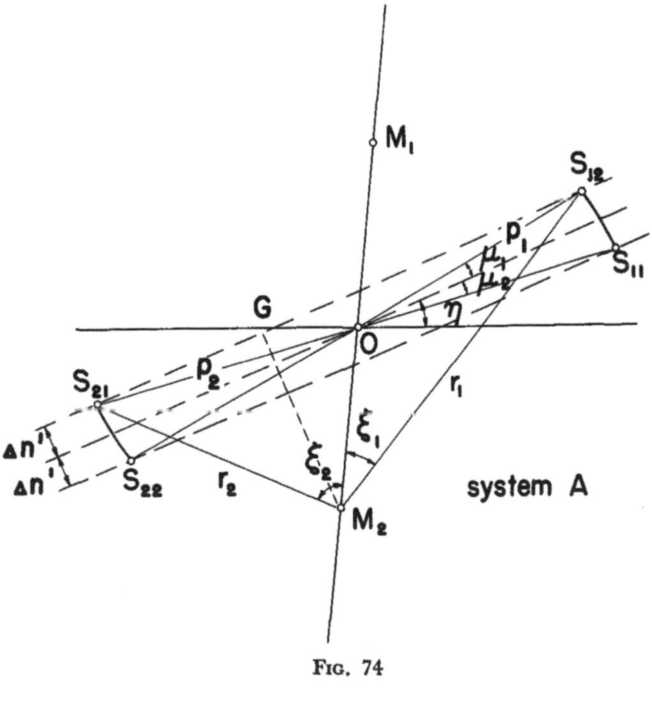
M₁
S₁₂
p₁
μ₁
μ₂
S₁₁
η
G
O
S₂₁
p₂
r₁
ξ₁
Δn'
Δn'
ξ₂
S₂₂
r₂
system A
M₂

FIG. 74

undesired motions is required. In this case any disturbance will be reduced rapidly to a periodic motion of very small amplitude and high frequency around the regular path.

The influence of the neutral zone imperfection (discussed in Sec. 6.3) is quite different from the other two types. A control mechanism having only this type of imperfection would possess all of the unfavorable characteristic qualities of ideal control. Therefore, neutral zones (such as those produced by backlash, for example) must be eliminated. Or they may be made ineffective by deliberately introducing a time lag of proper size into the control mechanism (see Fig. 67).

The theory indicates simple experiments which enable the manufacturer to determine the sizes of the various imperfections (see pp. 91 and 98) in existing control systems by measuring the frequency of the periodic motions.

7. EFFECTS UPON MOTION WITH VELOCITY CONTROL

Three different types of imperfections which may occur in a mechanism with velocity control are described as follows:

(1) The control element changes its velocity, not at $F = 0$, but when $|F|$ has reached a value $|\sigma_2|$ after passing zero (threshold imperfection). See Fig. 75a:

$$\left.\begin{aligned} &\dot{\beta} \text{ changes if } F = -\sigma_2 \text{ for } \dot{F} < 0 \\ &\dot{\beta} \text{ changes if } F = +\sigma_2 \text{ for } \dot{F} > 0 \end{aligned}\right\} \tag{147}$$

(2) The control element changes its velocity at constant time t_r after F passes zero (direct time lag imperfection). See Fig. 75b.

(3) The control element has no velocity in maintaining its position during an interval of time where $|F|$ is smaller than $|\sigma_1|$ (neutral zone imperfection; see Fig. 75c). This neutral zone $2\sigma_1$ has a "dead time" of variable size.

System C_1	*System* C_2
$F > \sigma_1, \quad \dot{\beta} = -V$	$F > \sigma_1, \quad \dot{\beta} = +V$
$\sigma_1 \geqslant F \geqslant -\sigma_1 \begin{cases} \dot{\beta} = 0 \\ \beta = \text{constant} \end{cases}$	$\sigma_1 \geqslant F \geqslant -\sigma_1 \begin{cases} \dot{\beta} = 0 \\ \beta = \text{constant} \end{cases}$
$F < -\sigma_1, \quad \dot{\beta} = +V$	$F < \sigma_1, \quad \dot{\beta} = -V$ (148)

In cases (1) and (2) the function $\beta(t)$ is a "saw-tooth" line. In case (3) the function is again of "saw-tooth" form but with flat tops replacing the peaks. The Figs. 75a, b, c, correspond to "model examples" (see below and Figs. 76, 80, 82) and will be discussed more thoroughly later.

All three types of imperfections will disturb the coincidence of the zeros of F with the switching of the relays. Thus our main task will be to study the location of the switch points with respect to the points where $F = 0$. To demonstrate the effects of the imperfections, we shall take a special example treated earlier in Sec. 5.2 (Example IV). At that time we explained briefly the effect of a small time lag. In this chapter the influence of all three types of imperfections on this "model" example will be demonstrated.

Let us restrict the discussion to control with feedback, which has been proved more efficient than that without feedback (Sec. 5.5).

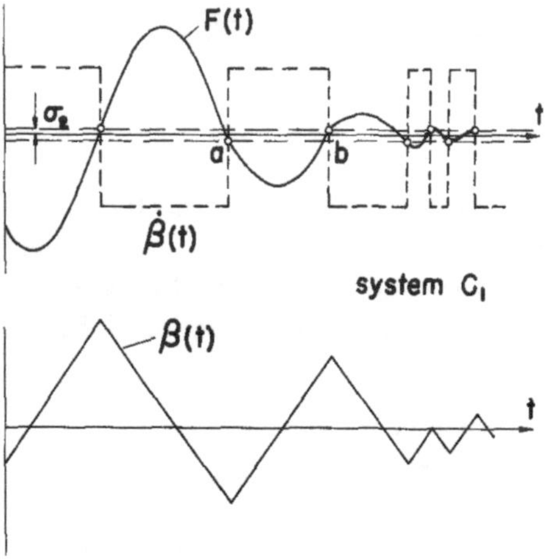

FIG. 75a

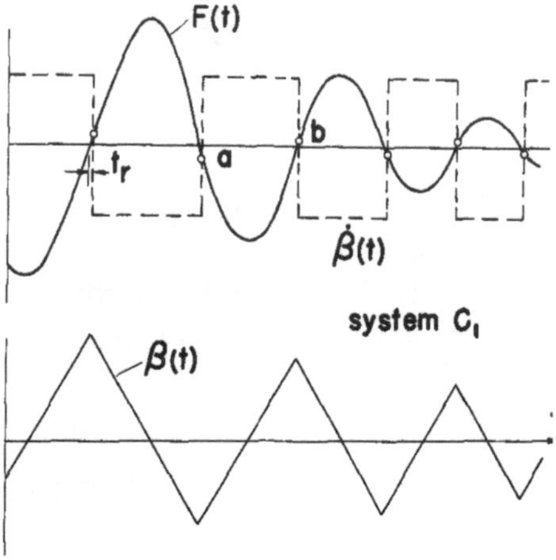

FIG. 75b

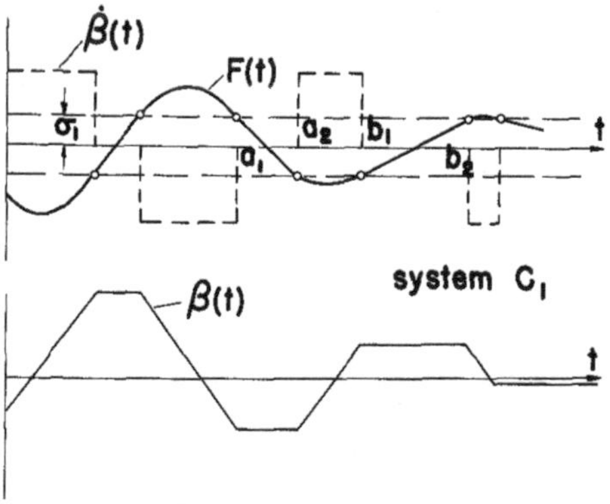

FIG. 75c

7.1. Constant lag dependent upon the value of the control function (threshold). Conditions for the switch points are the following:

$$\left.\begin{aligned} F &= -\sigma_2 \left\{\begin{matrix}\text{for type } a \text{ in system } C_1 \\ \\ \text{for type } b \text{ in system } C_2\end{matrix}\right\} \operatorname{sgn} F' < 0 \\ F &= +\sigma_2 \left\{\begin{matrix}\text{for type } b \text{ in system } C_1 \\ \\ \text{for type } a \text{ in system } C_2\end{matrix}\right\} \operatorname{sgn} F' > 0 \end{aligned}\right\} \quad (149)$$

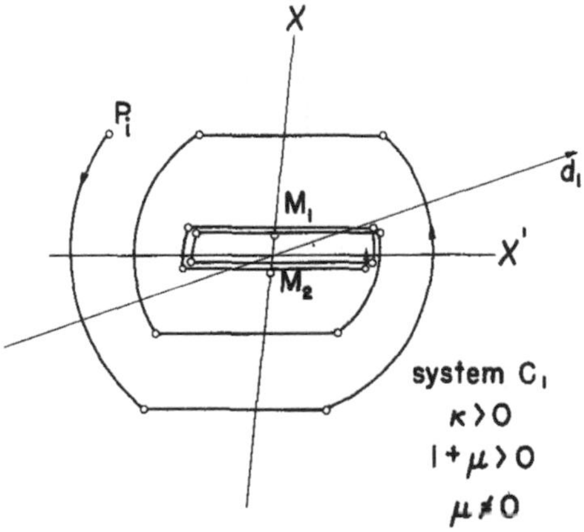

FIG. 76a

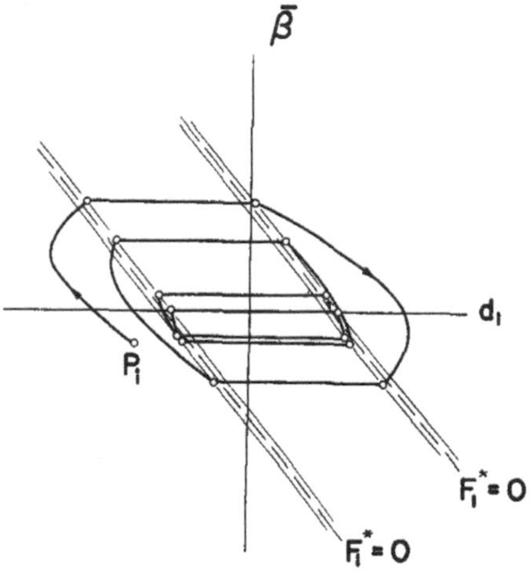

FIG. 76b

If F_1^* is used instead of F (see Sec. 3.2) then σ_2 is replaced by

$$\sigma_2^* = \sigma_2 \frac{\omega^2}{N} \frac{\omega}{V} \frac{1}{\rho_1} \quad (150)$$

Because $F_1^* =$ constant describes planes parallel to the planes $F_1^* = 0$, it is obvious that the switching planes will be parallel to the planes $F_1^* = 0$. The distance between the planes $F_1^* = 0$ and $F_1^* =$ const. will be given by (look at the projection in the β, d_1 plane)

$$\Delta n = \frac{\sigma_2^*}{1 + \mu} \cos \delta \quad (151)$$

with δ given by equation (102), or

$$\Delta n = \sigma_2^* \frac{\sqrt{1 - D^2}}{\sqrt{(1 + \mu)^2(1 - D^2) + (1 - 2\kappa D + \kappa^2)}} \quad (152)$$

The influence of a threshold imperfection on our model example (Sec. 5.2, Example IV, system C_1, $\kappa > 0$, $1 + \mu > 0$) is shown in Fig. 76a, b.

So long as the motion is not close to the region in which ideal control end points occur, the influence of the imperfection is almost negligible. In that special region the motion is appreciably different because the imperfection allows the motion to continue. As in the case of position control, this type of imperfection will remove all of the difficulties which are produced in the ideal system by the discontinuity of $F_1^{*\prime}$ at the switch points. Fig. 77 compares

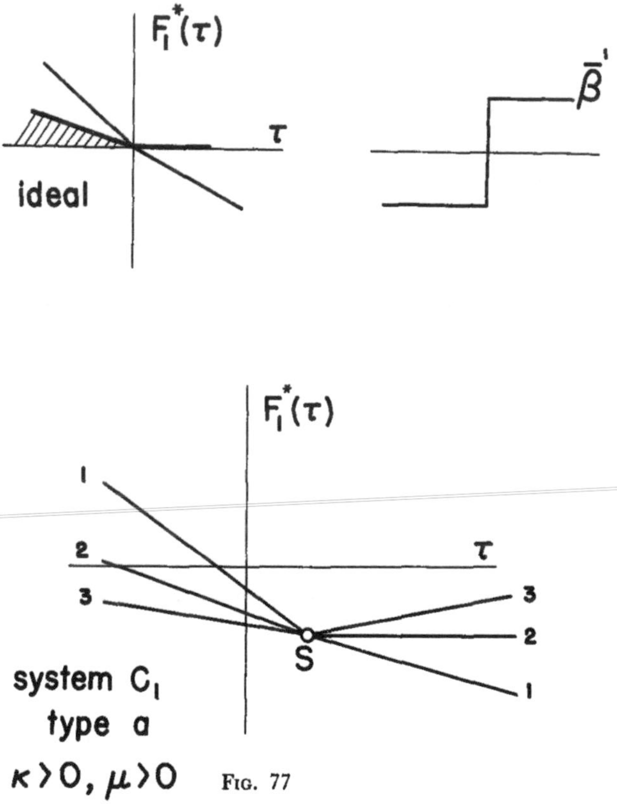

FIG. 77

the diffraction law for an ideal system and a system having a threshold imperfection for a switch point of type a with system C_1, $\kappa > 0$. It is obvious that a motion leading to S from any direction may continue beyond that point.

If the purpose of the control is to reduce a disturbance to zero rapidly, it is advisable to choose κ and μ of the control mechanism such that for ideal control end points would occur ($\mu \neq 0$; see Sec. 5.5) and to keep the imperfection σ_2^* small. The size of this imperfection will determine the length of the period and the amplitude of the periodic motion which represents the asymptotic state of motion in this case.

The period of this asymptotic motion may be found by introducing the switch point coordinates of the possible periodic motion into the equation

$$F_1^* = \sigma_2^* \operatorname{sgn} F_1^* \tag{153}$$

For example in system C_1 with a switch point of type a,

$$\bar{\psi}_p + \kappa\bar{\psi}_p' + \mu\bar{\beta}_p = -\sigma_2^* \tag{154}$$

The coordinates of those switch points are given by equations (120) in the χ, χ', $\bar{\beta}$ system.

A comparison of Figs. 76 and 78 shows that only for small σ_2^* is the asymptotic state a high-frequency periodic motion. If σ_2^* happens to be large (Fig. 78), the asymptotic motion is periodic, but not of high frequency.

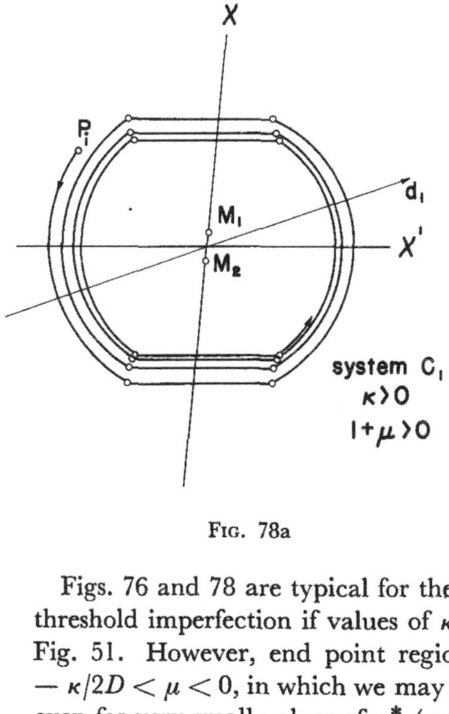

FIG. 78a

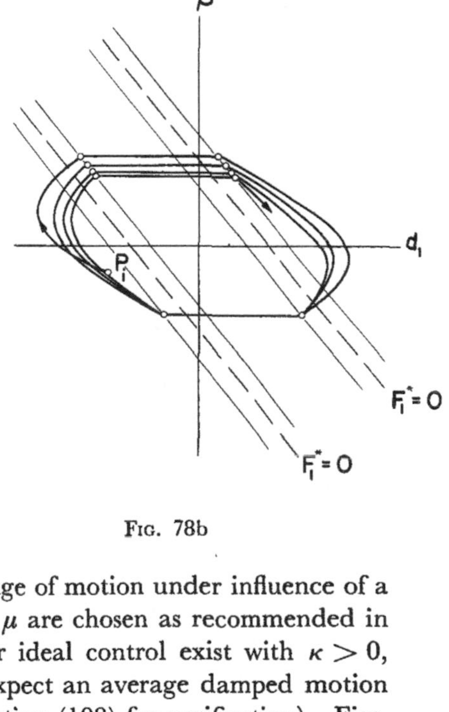

FIG. 78b

Figs. 76 and 78 are typical for the change of motion under influence of a threshold imperfection if values of κ and μ are chosen as recommended in Fig. 51. However, end point regions for ideal control exist with $\kappa > 0$, $-\kappa/2D < \mu < 0$, in which we may not expect an average damped motion even for very small values of σ_2^* (see equation (108) for verification). Figs. 79a-d show what types of motion may occur here. For the case $D = 0.1$, $\kappa = 2$, $\mu = -2$ (Fig. 79a, b) the motion tends toward a periodic one with low frequency. For the case $D = 0.1$, $\kappa = \frac{1}{2}$, $\mu = -2$ (Fig. 79c, d) the control will soon go to rest and the amplitude will grow ($\chi \to -2D$, $\chi' \to 0$, $\bar{\beta} \to +\infty$; or $\bar{\psi} \to +\infty$, $\bar{\psi}' \to +1$, $\bar{\beta} \to \infty$).

For an ideal control mechanism we established charts of μ and κ (see Figs. 51) which gave a complete survey of the motions that could occur for various combinations of κ and μ. To avoid a space representation in κ, μ, and σ_2^* it would be feasible to establish such charts for a given constant

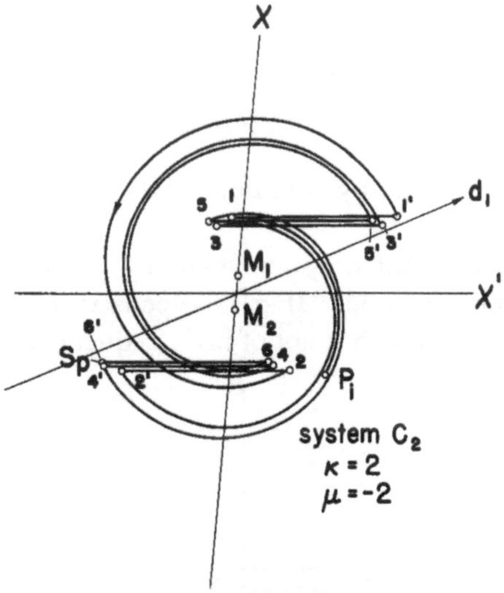

FIG. 79a

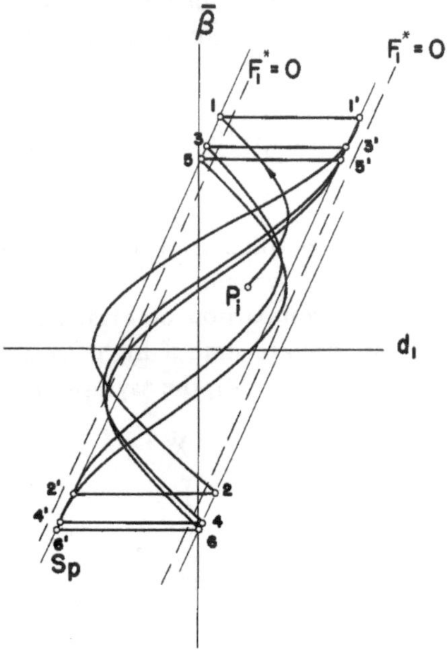

FIG. 79b

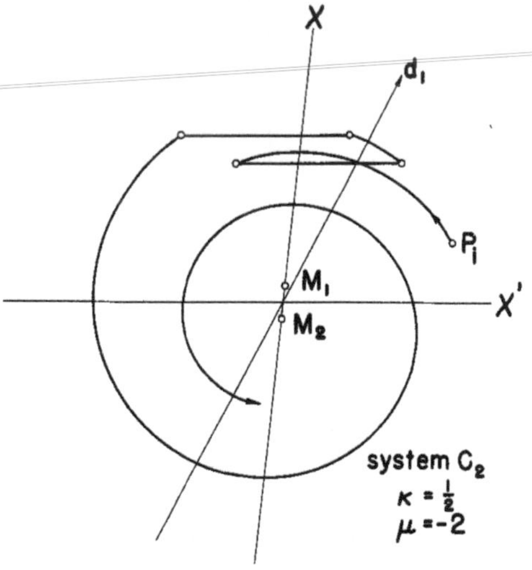

FIG. 79c

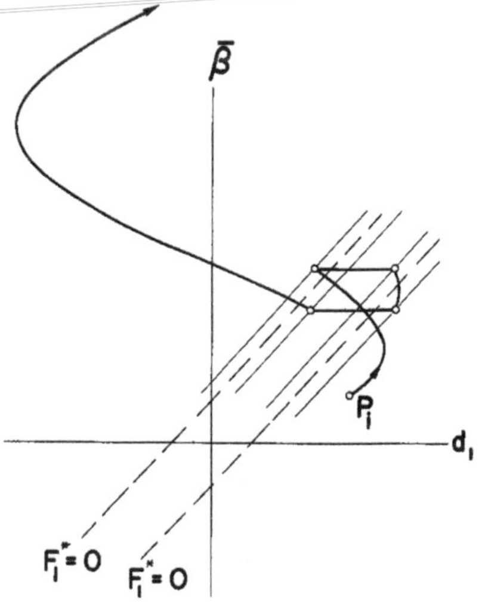

FIG. 79d

σ_2^* for a control mechanism having threshold imperfection. Such a chart for constant small σ_2^* would have slightly changed loci for the periodic motions because equation (122) would have to be replaced by

$$\kappa_p = -\frac{1}{\bar{\psi}_p^{*\prime}}\left[\chi_p^* + \bar{\beta}_p^* (1+\mu_p) \pm \sigma_2^*\right] \tag{155}$$

The positive sign of σ_2^* corresponds to system C_1 and the negative sign to system C_2. These new charts would have no regions of end points (or starting points). However, enlarged regions of periodic motions and rest points would appear.

7.2. Direct time lag. For a control system having direct time lag the switch points will again lie on planes in the χ, χ', $\bar{\beta}$ space. But these planes will *not* be parallel to the planes $F_1^* = 0$, as in the case of a threshold imperfection. A computation similar to that made for a time lag imperfection in the case of position control leads to the following equation for the locus of switch points:

$$\begin{aligned}&[\chi_s \mp 2D \operatorname{sgn}(-F_1^*)]\frac{\sin(\nu\Delta\tau + \sigma - \eta)}{\sin(\sigma-\eta)} e^{D\Delta\tau}\\ &+ \chi_s' \frac{\sin(\eta - \nu\Delta\tau)}{\sin(\sigma-\eta)} e^{D\Delta\tau} + (1+\mu)\bar{\beta}_s\\ &\mp \operatorname{sgn}(-F_1^*)[-2D + \kappa - (1+\mu)\Delta\tau] = 0\end{aligned} \tag{156}$$

For small values of $\Delta\tau$ this equation may be approximated by the form

$$\begin{aligned}&\chi_s(1+\kappa\Delta\tau) + \chi_s'[\kappa - (1-2\kappa D)\Delta\tau] + (1+\mu)\bar{\beta}_s\\ &\mp \operatorname{sgn}(-F_1^*)[\kappa(1+2D\Delta\tau) - (1+\mu)\Delta\tau] = 0\end{aligned} \tag{157}$$

From this equation it is easily seen that the angle which these switching planes form with the planes $F_1^* = 0$ depends upon $\Delta\tau$ and tends toward zero as $\Delta\tau \to 0$. Thus there exists an analogy to the influences of time lag on a position type control with regard to the location of the switching line. Also it is possible to compute the angles ζ and δ which correspond to the new positions of the switching planes (and also d_{00} and $\bar{\beta}_{00}$) and thus build new frames analogous to those of the ideal control. However, it is considerably more interesting to study the time lag influences in the diagrams of the ideal controlled motion. For this reason we shall retain our previous representation and construct the phase curve projections by using the fact that the time lag is apparent in the phase curve construction in the χ, χ' plane. Also, the fact that the change in $\bar{\beta}$ is proportional to τ will be utilized. Thus inclusion of the time lag in our previous construction is easily possible.

The switch points, when projected into the d_1, $\bar{\beta}$ plane, will not lie on straight lines because the switching planes are not parallel to the plane $F_1^* = 0$. Hence, the switching planes are not perpendicular to the d_1, $\bar{\beta}$ plane.

We have already studied briefly the effect of time lag on our model example (Sec. 5.2, Example IV, and Fig. 44). Here a motion occurs, one which under ideal control leads to an end point, but the imperfection allows the motion to continue beyond this undefined point. Fig. 80 shows the motion when time lag acts in the regions where its effect is not so important. The motion finally becomes a high frequency one around $\bar{\psi} = \chi + \bar{\beta} = 0$, $\bar{\psi}' = \chi' + \bar{\beta}' = 0$, $\bar{\beta} = 0$ in agreement with the former discussions (Sec. 5.2, Example IV, and Sec. 5.5). Substitution of the

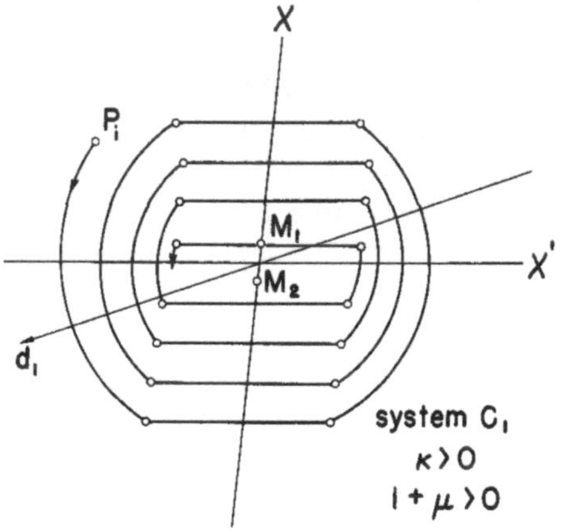

FIG. 80a

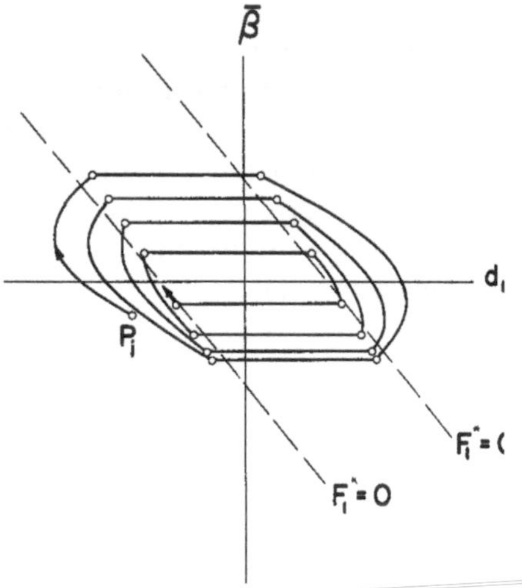

FIG. 80b

expressions for the periodic switch points (equation (120)) for small values of τ_p shows that the period of the high-frequency motion is given by

$$2\tau_p = 4\Delta\tau \tag{158}$$

Because of the presence of time lag, the switch points and the "commanding points" $F_1^* = 0$ do not coincide, and end points never occur. Thus we conclude that time lag imperfection acts essentially in the same manner as a delay dependent upon the value of F_1^* (threshold), as long as the influence of $\Delta\tau$ is not sufficient to change the sign of the essential coordinates of the switching planes δ, ζ, $\bar{\beta}_{00}$, and d_{00}. Of course their values vary slightly with $\Delta\tau$. Equation (157) shows that such sign changes may occur if κ is positive and very small and μ is positive and not too small. The coefficient of χ_s' and the constant term may change their sign with increasing $\Delta\tau$. Thus the switching planes might have entirely different positions for the cases of a finite $\Delta\tau$ and for $\Delta\tau = 0$.

It would appear to involve too many special details to develop all the formulas required to describe this foregoing discussion mathematically, but we may state the essential result adequately as follows: Time lag is favorable because end points are avoided, provided that values of κ and μ

are chosen to assure a good damping of the average motion (see equations 108, 125 and 126. As in the case of position control, the introduction of a negative time lag would have no advantage if a rapid damping of a disturbance is required.

7·3. Neutral Zone. As described at the beginning of this chapter the control element may have three different velocities: $\mathring{\beta} = -V, 0, +V$. In the intervals where $\mathring{\beta} = \mp V$, or $\bar{\beta}' = \mp \operatorname{sgn} F_1^*$, the motion will be described by equations (86) and (87). In the intervals with

$$-\sigma_1^* < [F_1^* = \bar{\psi} + \kappa\bar{\psi}' + \mu\bar{\beta}'] < +\sigma_1^* \tag{159}$$

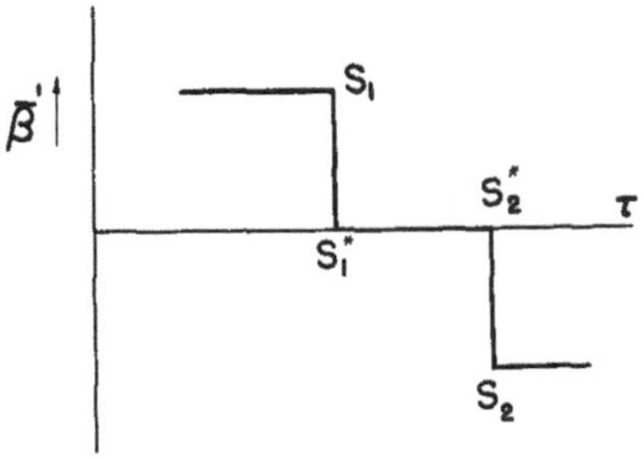

FIG. 81

the motion will be governed by the equation

$$\bar{\psi}_n'' + 2D\bar{\psi}_n' + \bar{\psi}_n = \bar{\beta}_{ni} = \text{const.} \tag{160}$$

with the solutions

$$\bar{\beta} = \bar{\beta}_{ni}$$

$$\left.\begin{aligned} \bar{\psi}_n(\tau) &= 2C_n e^{-D\tau} \cos(\nu\tau + \varepsilon_n) + \bar{\beta}_{ni} \\ \bar{\psi}_n'(\tau) &= 2C_n e^{-D\tau} \cos(\nu\tau + \varepsilon_n + \sigma) \end{aligned}\right\} \tag{161}$$

Thus we see that the construction of the phase curves is somewhat more complicated than for the other two types of imperfections. In the χ, χ' plane we must use spirals around $(\pm 2D, 0)$ as described earlier for the regular portion of the motion, and around (0, 0) for the neutral zone. In the d_1, $\bar{\beta}$ plane a damped cosine curve having as its neutral axis $d_1 = \pm 2D \sin(\sigma - \zeta)$ is used for the regular portion of the motion, and for the neutral zone the corresponding curve will be $\bar{\beta} = \text{const}$.

The transition from one interval to another must be done carefully, taking into account the facts that $\bar{\psi}$, $\bar{\psi}'$, and $\bar{\beta}$ are all continuous functions and that $\bar{\beta}'$ is discontinuous at the switch points. For example, Fig. 81 shows a switching from $\bar{\beta}' = +1$ to $\bar{\beta}' = 0$ and finally to $\bar{\beta}' = -1$. Using equations (91), (95), and (161) we find

$$\begin{aligned} \chi_{s_1} &= \chi_{s_1^*}, \quad \chi'_{s_1^*} = \chi'_{s_1} + 1 \\ \chi_{s_2^*} &= \chi_{s_2}, \quad \chi'_{s_2} = \chi'_{s_2^*} + 1 \end{aligned} \tag{162}$$

The "jump" of two units in the ideal control (equation (105)) is now split into two jumps of one unit each. The $\bar{\psi}$, $\bar{\psi}'$, $\bar{\beta}$ space is no longer divided into two half spaces, but is divided into three portions. Two of these are of the semi-infinite type already found, and the third is a "slice" of finite width depending in size upon σ_1^*.

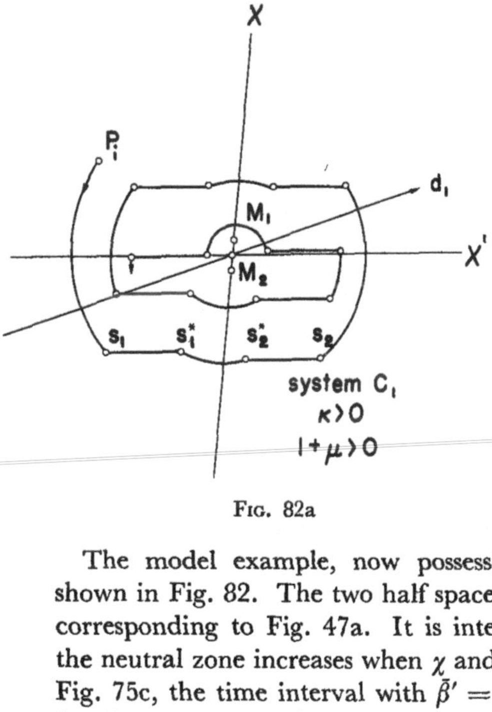

FIG. 82a

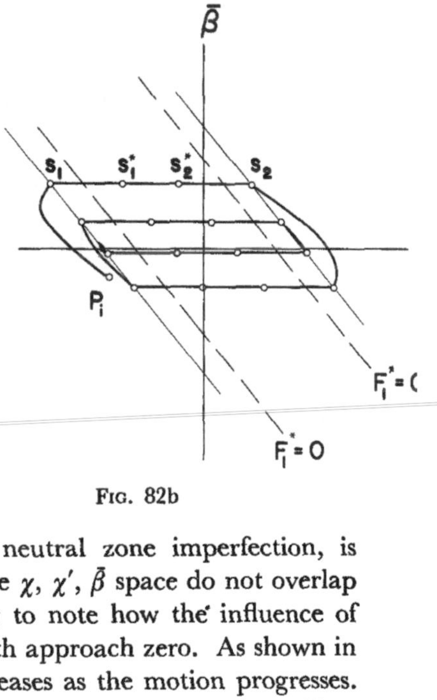

FIG. 82b

The model example, now possessing a neutral zone imperfection, is shown in Fig. 82. The two half spaces in the χ, χ', $\bar{\beta}$ space do not overlap corresponding to Fig. 47a. It is interesting to note how the influence of the neutral zone increases when χ and χ' both approach zero. As shown in Fig. 75c, the time interval with $\bar{\beta}' = 0$ increases as the motion progresses. In Fig. 83 an example is shown for which the half spaces do overlap. Note that all switch points lie on planes parallel to the plane $F_1^* = 0$.

In the case of a neutral zone imperfection the discontinuity of $F_1^{*\prime}$ at the switch points is given by (see Fig. 75c)

$$\left.\begin{aligned} F_1^{*\prime}(+0) - F_1^{*\prime}(-0) &= \mu \text{ (switch points type } a_1 \text{ and } a_2) \\ F_1^{*\prime}(+0) - F_1^{*\prime}(-0) &= -\mu \text{(switch points type } b_1 \text{ and } b_2) \end{aligned}\right\} \tag{163}$$

For the threshold and time lag imperfections we found that the difference between the commanding point $F_1^* = 0$ and the actual switching point denied the occurrence of end points. However, with the neutral zone imperfection, switching occurs at the commanding point. A figure analogous to Fig. 64a for position control would show this immediately, the only difference being that the discontinuity of the slope depends upon μ and not upon κ. Therefore, in a system with neutral zone, starting points and end points may occur. A control with neutral zone imperfection has all of the disadvantages of an ideal control. If the neutral zone imperfection is combined with a threshold or a time lag imperfection, these disadvantages will disappear.

It would be possible to construct charts analogous to Fig. 51 for constant neutral zone σ_1^*. As earlier, the values of μ and κ would essentially deter-

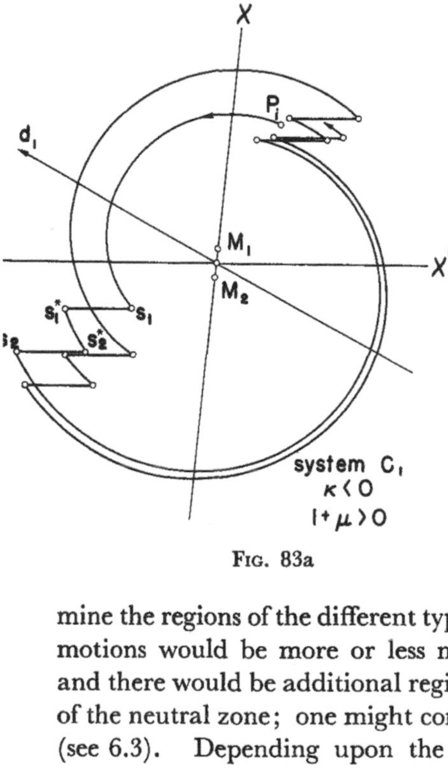

FIG. 83a

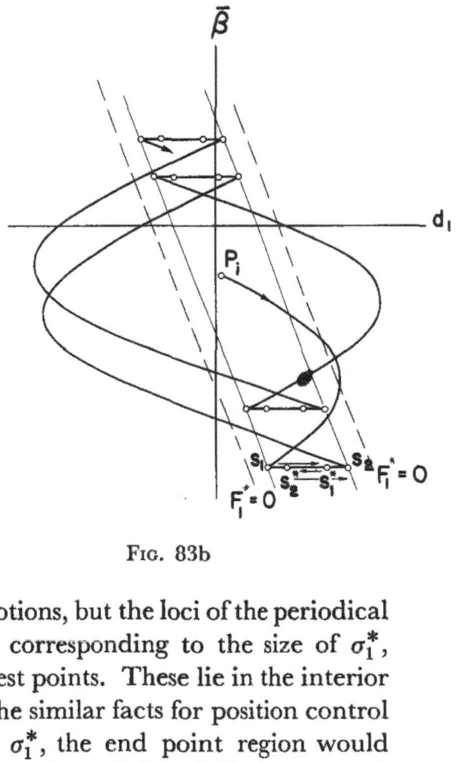

FIG. 83b

mine the regions of the different types of motions, but the loci of the periodical motions would be more or less modified corresponding to the size of σ_1^*, and there would be additional regions of rest points. These lie in the interior of the neutral zone; one might compare the similar facts for position control (see 6.3). Depending upon the size of σ_1^*, the end point region would also be more or less modified, the switching planes being shifted compared to those for ideal control.

7.4. Importance of these imperfections in systems with velocity control. Imperfections of the threshold or time lag type are favorable if the coefficients κ and μ are chosen so that end points would occur in the ideal control, provided that the imperfections are small. These regions of end points are shown in the κ, μ charts (Fig. 51). If κ and μ are chosen in the shaded portions of these regions, the undesired disturbances will disappear rapidly under this type of control. The imperfection "neutral zone," which might easily occur should be kept small by every available means. Even then its adverse effect should be removed by superposition of a suitable time lag.

PART III

DISCONTINUOUS CONTROL OF A MOVING BODY WITH MORE THAN ONE DEGREE OF FREEDOM

In the preceding chapters we have studied the control of bodies of one degree of freedom. There it was found possible to give suggestions for choosing the free constants of the control mechanism in order to obtain the best results for rapid damping of undesired disturbances from a predetermined path. Frequently bodies have more than one degree of freedom, for example airplanes or missiles have six. Fortunately, in most of these cases it is possible to assume that the longitudinal and lateral motions may be treated independently. Then we may study two systems, each having three degrees of freedom.

In general, an oscillating system of three degrees of freedom has more than one natural frequency. Hence all computations and graphical constructions will be much more complicated than for a system having a single degree of freedom. A common procedure is to obtain the most important frequency and then to approximate the general motion by a simpler one having only this frequency. An unavoidable consequence of this analysis is the impossibility of satisfying all the initial conditions of the motion. Owing to the nonlinearity of discontinuous control problems, initial conditions play a very important role. Thus it seems advisable to avoid oversimplification.

Phase plane construction contributed profitably to the study of the controlled motion of a body with a single degree of freedom. For either position or velocity control we found that the motion of the body could be described easily by its phase curve. For motions utilizing position control this was a curve in the ψ, ψ' phase plane; for motions with velocity control this was a curve in the ψ, ψ', β phase space. In the latter case the type of control added a second degree of freedom; fortunately no second natural frequency appeared. Therefore, as the basis of all graphical constructions, we were able to employ a single logarithmic spiral. This is impossible in the present problem. Nevertheless, the knowledge gained in the earlier investigations will serve as a guide.

The longitudinal motion of a missile under position control will be studied in detail. In this case only two natural frequencies occur. With the help of two spirals corresponding to these two frequencies the essential features of the motion and its phase curve may be studied graphically. End points and rest points for the control will appear, and their location in the phase space will be discussed. Periodic motions will occur as possible asymptotic states of motion. Control mechanism imperfections are as important as in the systems with a single degree of freedom. They cause end points to disappear. However, there will be interesting differences in the details.

8. THE LONGITUDINAL MOTION OF A MISSILE

(Discontinuous Position Control)

8.1. The equations of motion, the control equation, and their formal solutions. The desired path of the missile should be a straight one, characterized by the velocity V_0, the flight path angle γ_0, and the angle

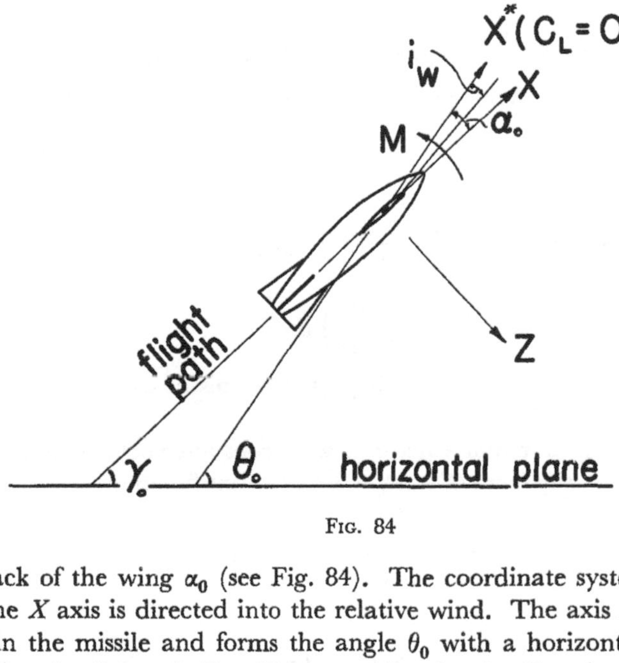

FIG. 84

of attack of the wing α_0 (see Fig. 84). The coordinate system is chosen so that the X axis is directed into the relative wind. The axis X^* ($C_L = 0$) is fixed in the missile and forms the angle θ_0 with a horizontal plane. This desired path of the missile will be considered to be disturbed in such a way that the resulting motion is symmetrical; that is, no lateral forces occur.

The deviations of the variables from their desired values are correspondingly named v, γ, α, and θ. The motion is described by two equations stating the equilibrium of forces in the X and Z directions and by one pitching moment equilibrium equation about the Y axis.

For the convenience of the reader we will write the equations of motion in detail. The notation, which corresponds to that commonly used by the National Advisory Committee for Aeronautics, is given in Table VII.[1]

[1] See Appendix I.

Also, numerical values that apply to the particular missile which is used as an example in later parts are shown in Table VIII.

The symmetrical motion of a missile without control is governed by the following equations:[1]

Equilibrium of forces in the X direction:

$$(C'_{D_0} - C_{L_0})\,\alpha + (C_W \cos\gamma_0)\,\theta + \left(2C^*_{D_0} + \frac{d}{d\tau}\right)\frac{v}{V_0} = 0$$

Equilibrium of forces in the Z direction:

$$\left(C'_{L_0} + C_{D_0} + \frac{d}{d\tau}\right)\alpha + \left(C_W \sin\gamma_0 - \frac{d}{d\tau}\right)\theta + 2C^*_{L_0}\frac{v}{V_0} = 0 \qquad (164)$$

Equilibrium of moments about the Y axis

$$\left(\bar{\mu}C'_{M_0} + C_{M^\circ_\alpha}\frac{d}{d\tau}\right)\alpha + \left(C_{M_q}\frac{d}{d\tau} - \frac{k_y^2}{c^2}\frac{d^2}{d\tau^2}\right)\theta = 0$$

For ease of writing the coefficients in equations (164) may be abbreviated as:

$$\begin{aligned}
&a_{10}\alpha + b_{10}\theta + \left(c_{10} + c_{11}\frac{d}{d\tau}\right)\frac{v}{V_0} = 0\\
&\left(a_{20} + a_{21}\frac{d}{d\tau}\right)\alpha + \left(b_{20} + b_{21}\frac{d}{d\tau}\right)\theta + c_{20}\frac{v}{V_0} = 0 \qquad (165)\\
&\left(a_{30} + a_{31}\frac{d}{d\tau}\right)\alpha + \left(b_{31}\frac{d}{d\tau} + b_{32}\frac{d^2}{d\tau^2}\right)\theta = 0
\end{aligned}$$

The values of a_{mn}, b_{mn}, and c_{mn} for the particular missile used in the examples are given in Table VIII.

This system of differential equations with constant coefficients has well-known solutions in the form of sums of exponential functions:

$$\begin{aligned}
\theta &= \sum_{k=1}^{4} M_k e^{\lambda_k\tau}\\
\frac{v}{V_0} &= \sum_{k=1}^{4} C_k M_k e^{\lambda_k\tau} \qquad (166)\\
\alpha &= \sum_{k=1}^{4} A_k M_k e^{\lambda_k\tau}
\end{aligned}$$

[1] These equations correspond to those found in Division N, Dynamics of the Airplane, by B. M. Jones, *Aerodynamic Theory*, Vol. V, W. F. Durand, Ed., Calif. Inst. of Techn., 1943, p. 171. Jones uses different notations (explained in Table 1 and 2 on p. 133 of his article). It is obvious that $\left(\frac{w}{V}\right)_{\text{Jones}} = \alpha$, and $\left(\frac{u}{V}\right)_{\text{Jones}} = \frac{v}{V_0}$ in our notation. The reader should be reminded too that in the reference, forces are related to $\rho V_0^2 S$ instead of $1/2\rho V_0^2 S$, as is customary. Thus $k_L = (1/2)C_L$.

The λ_k are the roots of the characteristic determinant of the system:

$$D = \begin{vmatrix} a_{10} & b_{10} & c_{10} + c_{11}\lambda \\ a_{20} + a_{21}\lambda & b_{20} + b_{21}\lambda & c_{20} \\ a_{30} + a_{31}\lambda & b_{31}\lambda + b_{32}\lambda^2 & 0 \end{vmatrix} \tag{167a}$$

or

$$r_4\lambda^4 + r_3\lambda^3 + r_2\lambda^2 + r_1\lambda + r_0 = 0 \tag{167b}$$

where r_k are constants dependent upon a_{mn}, b_{mn}, c_{mn}. Since $D = 0$ is an equation of fourth order, there are only four roots for λ.[1] If the uncontrolled missile is dynamically stable, all roots λ must have a negative real part. In this case the ratios r_3/r_4, r_2/r_4, r_1/r_4 and r_0/r_4 are positive, and

$$r_3r_2r_1 - r_4r_1^2 - r_0r_3^2 > 0.$$

The constants M_k are determined by the initial conditions $v(0)/V_0$, $\alpha(0)$, $\theta(0)$, and $\theta'(0)$. The C_k and A_k are determined by the coefficients of the system given in Table VIII.

The longitudinal motion of the missile may be controlled either by the ailerons or by the horizontal tailplane. The control surface deflection will be determined by the missile deviations. The remaining question is which of the deviations to be used for correcting the disturbed motion is best detected by instruments. A record of θ and $\theta' = d\theta/d\tau$ is easily obtained by gyroscopic measurements. Records of α and v/V_0 are more difficult to obtain and are less accurate. Therefore, we will study a control system depending on θ and θ'. In particular the angle of the control surface will depend on the sign of a linear function F of θ and θ': $F = \theta + \kappa\theta'$. In other words we have chosen a position control.

In the case of aileron control we may assume that only a force in the Z direction is applied and that the equilibrium of momentum is not affected. Introducing p_{mn} for the coefficients of α, θ, v/V_0 into the equations of motion (for example $p_{31} = a_{30} + a_{31}\, d/d\tau$), we have then for aileron control the following system instead of (165)

$$\left.\begin{aligned} p_{11}\alpha + p_{12}\theta + p_{13}\frac{v}{V_0} &= 0 \\ p_{21}\alpha + p_{22}\theta + p_{23}\frac{v}{V_0} &= -N_2\eta \\ p_{31}\alpha + p_{32}\theta &= 0 \end{aligned}\right\} L_2 \tag{168a}$$

with

$$\eta = \pm\eta_0 \operatorname{sgn}(\theta + \kappa\theta') = \pm\eta_0 \operatorname{sgn} F \tag{168b}$$

The two signs before $\eta_0 \operatorname{sgn} F$ again indicate that there are two essentially different control systems. It will be practical—for a close connection to

[1] A system with three degrees of freedom may lead to a characteristic determinant of sixth order at most.

the earlier work—to call that system in which the force applied by the control element is *added* to the existing restoring forces system A (see p. 130).

In the case of horizontal tailplane control it may be assumed that only the equilibrium of momentum is influenced and that the effect on the balance of forces is negligible.

Corresponding to equations (168) we obtain:

$$\left.\begin{aligned} p_{11}\alpha + p_{12}\theta + p_{13}\frac{v}{V_0} &= 0 \\ p_{21}\alpha + p_{22}\theta + p_{23}\frac{v}{V_0} &= 0 \\ p_{31}\alpha + p_{32}\theta &= N_3\eta_h \end{aligned}\right\} L_3 \qquad (169a)$$

with

$$\eta_h = \pm\eta_{h0}\,\mathrm{sgn}\,(\theta + \kappa\theta') = \pm\eta_{h0}\,\mathrm{sgn}\,F. \qquad (169b)$$

Both L_2 and L_3 systems remain linear between two consecutive switching points of the control system. The solutions are sums of exponential functions in every interval m between two switch points, but there will be additional constant terms.

$$\left.\begin{aligned} \theta_m &= \sum_{k=1}^{4} M_{k_m} e^{\lambda_k\tau} + B \\ \frac{v_m}{V_0} &= \sum_{k=1}^{4} C_{k_m} M_{k_m} e^{\lambda_k\tau} + C \\ \alpha_m &= \sum_{k=1}^{4} A_{k_m} M_{k_m} e^{\lambda_k\tau} + A \end{aligned}\right\} \qquad (170)$$

These constants A, B, and C are given by Table IX.

Table IX

L_2: Control by aileron, equations (168)	L_3: Control by horizontal tailplane, equations (169)
$A = 0$	$A = \frac{1}{a_{30}} N_3(\pm\,\eta_{h0}\,\mathrm{sgn}\,F)$
$B = \frac{c_{10}}{b_{10}c_{20} - c_{10}b_{20}} N_2\,(\pm\eta_0\,\mathrm{sgn}\,F)$	$B = \frac{c_{10}a_{20}}{a_{30}(b_{10}c_{20} - c_{10}b_{20})} N_3(\pm\eta_{h0}\,\mathrm{sgn}\,F)$
$C = \frac{-b_{10}}{b_{10}c_{20} - c_{10}b_{20}} N_2\,(\pm\eta_0\,\mathrm{sgn}\,F)$	$C = \frac{a_{10}b_{20} - b_{10}a_{20}}{a_{30}(b_{10}c_{20} - c_{10}b_{20})} N_3(\pm\eta_{h0}\,\mathrm{sgn}\,F)$

In the first interval the M_{k_1} are given by the initial values of the variables $\theta(0)$, $\theta'(0)$, $\alpha(0)$, and $v(0)/V_0$. In any other interval the M_{k_m} are given by the fact that θ, θ', α, and v/V_0 values at the beginning of the mth interval equal the $\theta, \theta', \alpha, v/V_0$ values at the end of the $(m-1)$th interval, for these functions are continuous.

The simplest way to get a form of the solutions (170) containing the initial conditions explicitly is to apply the Laplace transformation to equations (168) or (169). Some new notations will be helpful in writing down these solutions. Let p_{mn_k} designate the coefficient p_{mn} for the exponential function $e^{\lambda_k \tau}$.

$$p_{31} = a_{30} + a_{31}\frac{d}{d\tau}, \qquad p_{32} = b_{31}\frac{d}{d\tau} + b_{32}\frac{d^2}{d\tau^2}$$

$$p_{31_k} = a_{30} + a_{31}\lambda_k, \qquad p_{32_k} = b_{31}\lambda_k + b_{32}\lambda_k^2 \tag{171a}$$

Further

$$\Delta'(\lambda_k) = (\lambda_k - \lambda_1)(\lambda_k - \lambda_2)\ldots(\lambda_k - \lambda_{k-1})(\lambda_k - \lambda_{k+1})\ldots(\lambda_k - \lambda_n) \tag{171b}$$

For example

$$\Delta'(\lambda_2) = (\lambda_2 - \lambda_1)(\lambda_2 - \lambda_3)(\lambda_2 - \lambda_4)$$

With these notations the new form of the solutions (170) is

$$\theta_m - B = \sum_{k=1}^{4} \frac{e^{\lambda_k \tau}}{\Delta'(\lambda_k)} \begin{vmatrix} p_{11_k} & c_{11}\left(\dfrac{v_{im}}{V_0} - C\right) & p_{13_k} \\ p_{21_k} & a_{21}(\alpha_{i_m} - A) + b_{21}(\theta_{i_m} - B) & p_{23_k} \\ p_{31_k} & \left\{\begin{matrix} a_{31}(\alpha_{i_m} - A) \\ + (b_{32}\lambda_k + b_{31})(\theta_{i_m} - B) \\ + b_{32}\theta'_{i_m} \end{matrix}\right\} & p_{33_k} \end{vmatrix} \tag{172a}$$

$$\alpha_m - A = \sum_{k=1}^{4} \frac{e^{\lambda_k \tau}}{\Delta'(\lambda_k)} \begin{vmatrix} c_{11}\left(\dfrac{v_{im}}{V_0} - C\right) & p_{12_k} & p_{13_k} \\ a_{21}(\alpha_{i_m} - A) + b_{21}(\theta_{i_m} - B) & p_{22_k} & p_{23_k} \\ \left\{\begin{matrix} a_{31}(\alpha_{i_m} - A) \\ + (b_{32}\lambda_k + b_{31})(\theta_{i_m} - B) \\ + b_{32}\theta'_{i_m} \end{matrix}\right\} & p_{32_k} & p_{33_k} \end{vmatrix} \tag{172b}$$

$$\frac{v_m}{V_0} - C = \sum_{k=1}^{4} \frac{e^{\lambda_k \tau}}{\Delta'(\lambda_k)} \begin{vmatrix} p_{11_k} & p_{12_k} & c_{11}\left(\dfrac{v_{im}}{V_0} - C\right) \\ p_{21_k} & p_{22_k} & a_{23}(\alpha_{i_m} - A) + b_{21}(\theta_i - B) \\ p_{31_k} & p_{32_k} & \left\{\begin{matrix} a_{31}(\alpha_{i_m} - A) \\ + (b_{32}\lambda_k + b_{31})(\theta_{i_m} - B) \\ + b_{32}\theta'_{i_m} \end{matrix}\right\} \end{vmatrix} \tag{172c}$$

We have yet to decide which sign before η_0 or η_{h0} corresponds to control system A (see p. 128); we may do so now (compare with Sec. 2.1). The control function $F = \theta + \kappa\theta'$ will vary similarly to θ for small κ. θ is determined by a fourth order differential equation which may replace the L_2 or L_3 system.

$$r_4\theta^{IV} + r_3\theta''' + r_2\theta'' + r_1\theta' + r_0(\theta - B) = 0.$$

For a stable missile without control $r_0/r_4 > 0$ (see p. 127). If we assume this basic stability to exist, then a negative value of B for positive θ would increase the term $(\theta - B)$ and thus reinforce the restoring force. Hence, we find that if

$$\frac{c_{10}}{b_{10}c_{20} - c_{10}b_{20}} > 0$$

the lower sign before η_0 would designate control system A in L_2. But because $a_{20}/a_{30} < 0$, the upper sign before η_{h0} would designate control system A in L_3.

8.2. Continuity qualities of the solutions. The functions v/V_0, α, θ and θ' will be continuous because they are continuous in any interval and their values at the end of any interval are equal to the initial values for that following. In system L_2 the function $-N_2\eta$ is discontinuous, and similarly in system L_3 the function $N_3\eta_h$ is discontinuous. With respect to the continuity of v/V_0, α, θ, and θ' we can easily determine the continuity qualities of the solutions in general. In both cases the control function will have a discontinuous slope at the switch points of the phase space.

Table X

Continuity quality	Function in system L_2				Function in system L_3			
Smooth	$\frac{v}{V_0}$		θ		$\left(\frac{v}{V_0}\right)'$	α	θ	
Broken line	$\left(\frac{v}{V_0}\right)'$	α	θ'	$F = \theta + \kappa\theta'$		α'	θ'	$F = \theta + \kappa\theta'$
Step		α'	θ''	$F' = \theta' + \kappa\theta''$			θ''	$F' = \theta' + \kappa\theta''$

Fig. 85 shows the behavior of $\theta(\tau)$, $\theta'(\tau)$, $\alpha(\tau)$, and $v(\tau)/V_0$ in the neighborhood of a switch point. The portion shown is taken from the general figure describing the motion of a particular missile (the physical data of which are given in Table VIII) after a disturbance from straight forward flight.[1]

[1] The linearization of the flight equations sets certain limits for the maximum values of v/V_0, θ, θ' and α. For better visualization in this special example, and in some following ones, initial disturbances are chosen too large in that respect.

The aileron control is utilized in system L_2. The discontinuous slopes of α and θ' are very conspicuous.

As previously mentioned, the motion is composed of two superposed oscillations of very different frequencies (for the missile in question: $\nu_1 = 0.2777$ and $\nu_3 = 7.062$). It is obvious that the discontinuity of the

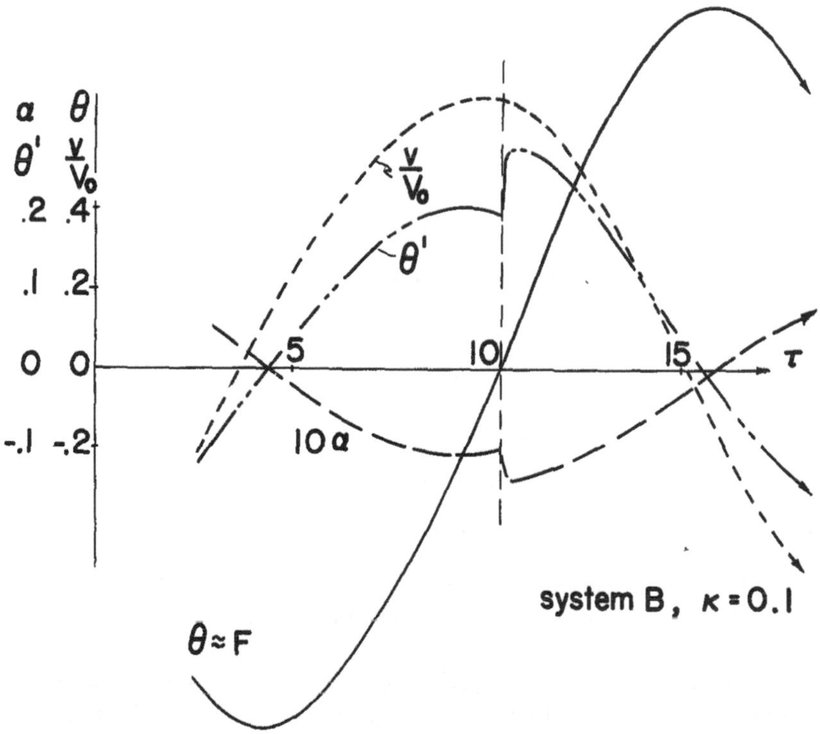

FIG. 85

slopes of α and θ' at the switch points is nicely smoothed by the oscillation of high frequency. Therefore neglecting this oscillation would seriously affect the validity of any computational results.

8.3. Representation of the motion in the phase space. The motion of the missile is predetermined by the four initial values θ_i, θ_i', α_i, v_i/V_0. Thus it may be described in a phase space of four dimensions, corresponding to the four variables θ, θ', α, v/V_0. The control function $F = \theta + \kappa\theta'$ would be a plane in such a generalized space. This representation is undesirable because there is no easy means available for construction of the phase curve in such a space.

A thorough investigation shows that the methods employed for constructing the phase curve of single degree of freedom systems may be applied here in a generalized form.

θ, α and v/V_0 are given by the equations (170) or (172) and θ' by

$$\theta' = \frac{d}{d\tau}(\theta - B) \tag{172d}$$

The constants A, B, C, are given by Table IX. The form of the solutions (172) shows how the amplitude of each oscillation depends on the initial conditions. Because the roots λ_1 and λ_2, λ_3 and λ_4, are complex conjugate, there are two different frequencies. Thus we may write (omitting the subscript m for convenience)

$$\theta - B = \xi_1 + \bar{\xi}_1 + \xi_3 + \bar{\xi}_3 \tag{173}$$

where (see equation (170))

$$\xi_k = M_k e^{\lambda_k \tau} \text{ and } \lambda_k = \delta_k + i\nu_k \tag{174}$$

The sum $\xi_k + \bar{\xi}_k$ will always be real and of the form

$$\xi_k + \bar{\xi}_k = 2|M_k| e^{\delta_k \tau} \cos(\nu_k \tau + \varepsilon_k) \tag{175a}$$

with

$$M_k = |M_k| e^{i\varepsilon_k} \tag{175b}$$

The derivative of the sum has the form

$$\xi_k' + \bar{\xi}_k' = 2|M_k| \cdot |\lambda_k| e^{\delta_k \tau} \cos(\nu_k \tau + \varepsilon_k + \sigma_k) \tag{175c}$$

with

$$\lambda_k = |\lambda_k| e^{i\sigma_k} \tag{175d}$$

Form (175c) is real also. Equations (175a) and (175c) remind us of the form of solutions obtained in the studies of systems with but one degree of freedom.

The representation of the motion of a missile in a phase space of four dimensions having coordinates θ, θ', α and v/V_0 would be difficult and could be managed only by employing many projections. It is suggested therefore that, instead of tracing all four variables in one space, we study the two different oscillations $(\xi_1 + \bar{\xi}_1)$ and $(\xi_3 + \bar{\xi}_3)$ separately. Thus working with two different spirals in respective phase planes, we might deduce all further details from them.

For phase plane construction we need the coordinate, in this case $\xi_k + \bar{\xi}_k$, and its time derivative. The dependence of these two "partial" phases on the variables θ, θ', α, and v/V_0 has now to be studied.

$$\theta - B = \xi_1 + \bar{\xi}_1 + \xi_3 + \bar{\xi}_3 \tag{176a}$$

$$\begin{aligned} \theta' &= \xi_1' + \bar{\xi}_1' + \xi_3' + \bar{\xi}_3' \\ &= \lambda_1 \xi_1 + \bar{\lambda}_1 \bar{\xi}_1 + \lambda_3 \xi_3 + \bar{\lambda}_3 \bar{\xi}_3 \end{aligned} \tag{176b}$$

$$\frac{v}{V_0} - C = C_1 \xi_1 + \bar{C}_1 \bar{\xi}_1 + C_3 \xi_3 + \bar{C}_3 \bar{\xi}_3 \tag{176c}$$

For reasons of symmetry it is advantageous to use v'/V_0 as the fourth equation instead of one for α. From equations (165, 168a and 169a) we find the relation between v'/V_0 and α to be

$$\frac{v'}{V_0} = -\frac{1}{c_{11}}\left(a_{10}\alpha + b_{10}\theta + c_{10}\frac{v}{V_0}\right)$$

Therefore, in addition to θ, θ', and v/V_0 we may use v'/V_0 instead of α.

$$\begin{aligned}\frac{v'}{V_0} &= C_1\,\xi_1' + \bar{C}_1\,\bar{\xi}_1' + C_3\,\xi_3' + \bar{C}_3\,\bar{\xi}_3' \\ &= C_1\,\lambda_1\,\xi_1 + \bar{C}_1\,\bar{\lambda}_1\,\bar{\xi}_1 + C_3\,\lambda_3\,\xi_3 + \bar{C}_3\,\bar{\lambda}_3\,\bar{\xi}_3 \qquad (176d)\end{aligned}$$

Now if

$$D = \begin{vmatrix} 1 & 1 & 1 & 1 \\ \lambda_1 & \bar{\lambda}_1 & \lambda_3 & \bar{\lambda}_3 \\ C_1 & \bar{C}_1 & C_3 & \bar{C}_3 \\ C_1\lambda_1 & \bar{C}_1\bar{\lambda}_1 & C_3\lambda_3 & \bar{C}_3\bar{\lambda}_3 \end{vmatrix} \qquad (177)$$

We have

$$\xi_1 = \frac{1}{D}\begin{vmatrix} \theta - B & 1 & 1 & 1 \\ \theta' & \bar{\lambda}_1 & \lambda_3 & \bar{\lambda}_3 \\ \dfrac{v}{V_0} - C & \bar{C}_1 & C_3 & \bar{C}_3 \\ \dfrac{v'}{V_0} & \bar{C}_1\bar{\lambda}_1 & C_3\lambda_3 & \bar{C}_3\bar{\lambda}_3 \end{vmatrix} \qquad (178a)$$

$$\xi_3 = \frac{1}{D}\begin{vmatrix} 1 & 1 & \theta - B & 1 \\ \lambda_1 & \bar{\lambda}_1 & \theta' & \bar{\lambda}_3 \\ C_1 & \bar{C}_1 & \dfrac{v}{V_0} - C & \bar{C}_3 \\ C_1\lambda_1 & \bar{C}_1\bar{\lambda}_1 & \dfrac{v'}{V_0} & \bar{C}_3\bar{\lambda}_3 \end{vmatrix} \qquad (178b)$$

Also

$$\xi_1' = \lambda_1\,\xi_1 \qquad (178c)$$

$$\xi_3' = \lambda_3\,\xi_3 \qquad (178d)$$

For clarity the C_k are given explicitly. They can easily be found by comparing equations (170) and (172).

$$C_k = -\frac{\begin{vmatrix} p_{11_k} & p_{12_k} \\ p_{21_k} & p_{22_k} \end{vmatrix}}{\begin{vmatrix} p_{11_k} & p_{13_k} \\ p_{21_k} & p_{23_k} \end{vmatrix}} = -\frac{\begin{vmatrix} p_{21_k} & p_{22_k} \\ p_{31_k} & p_{32_k} \end{vmatrix}}{\begin{vmatrix} p_{21_k} & p_{23_k} \\ p_{31_k} & p_{33_k} \end{vmatrix}} = -\frac{\begin{vmatrix} p_{11_k} & p_{12_k} \\ p_{31_k} & p_{32_k} \end{vmatrix}}{\begin{vmatrix} p_{11_k} & p_{13_k} \\ p_{31_k} & p_{33_k} \end{vmatrix}} \quad (179)$$

$\theta - B$ and $v/V_0 - C$ are discontinuous at every switch point because B and C change their sign (see Table IX). The jumps $|\Delta(\xi_1 + \bar{\xi}_1)|$ and $|\Delta(\xi_3 + \bar{\xi}_3)|$ are given by

$$|\Delta(\xi_1 + \bar{\xi}_1)| = \left| \frac{2}{D} \begin{vmatrix} -B & 0 & 1 & 1 \\ 0 & \bar{\lambda}_1 - \lambda_1 & \lambda_3 & \bar{\lambda}_3 \\ -C & \bar{C}_1 - C_1 & C_3 & \bar{C}_3 \\ 0 & \bar{C}_1\bar{\lambda}_1 - C_1\lambda_1 & C_3\lambda_3 & \bar{C}_3\bar{\lambda}_3 \end{vmatrix} \right| \quad (180a)$$

$$|\Delta(\xi_3 + \bar{\xi}_3)| = \left| \frac{2}{D} \begin{vmatrix} 1 & 1 & -B & 0 \\ \lambda_1 & \bar{\lambda}_1 & 0 & \bar{\lambda}_3 - \lambda_3 \\ C_1 & \bar{C}_1 & -C & \bar{C}_3 - C_3 \\ C_1\lambda_1 & \bar{C}_1\bar{\lambda}_1 & 0 & \bar{C}_3\bar{\lambda}_3 - C_3\lambda_3 \end{vmatrix} \right| \quad (180b)$$

The corresponding jumps in the derivatives are given by

$$\Delta(\xi_1' + \bar{\xi}_1') = \delta_1 \cdot \Delta(\xi_1 + \bar{\xi}_1) + i\nu_1 \cdot \Delta(\xi_1 - \bar{\xi}_1) \quad (181a)$$

$$\Delta(\xi_3' + \bar{\xi}_3') = \delta_3 \cdot \Delta(\xi_3 + \bar{\xi}_3) + i\nu_3 \cdot \Delta(\xi_3 - \bar{\xi}_3) \quad (181b)$$

The jumps in $(\xi_k + \bar{\xi}_k)$ and $(\xi_k' + \bar{\xi}_k')$ at the switch points depend only on the coefficients of the differential equations describing the motion of a given missile. They do not vary with the coefficient κ of the control function, but they do change with the maximum correcting force applied. This force depends on η_0 (or η_{0_h}).

The initial values of ξ_k are found by putting the initial values of θ, θ', v/V_0, v'/V_0 into equations (178a, b). In addition, they are explicitly given in equations (172) as the coefficients of $e^{\lambda_k \tau}$.

From equation (168b) the control function is given by

$$\begin{aligned} F = \theta + \kappa\theta' &= \sum_{k=1}^{4}(1 + \kappa\lambda_k)M_k e^{\lambda_k \tau} + B \\ &= \sum_{k=1}^{4}(\xi_k + \kappa\xi_k') + B \\ &= (\xi_1 + \bar{\xi}_1) + \kappa(\xi_1' + \bar{\xi}_1') + (\xi_3 + \bar{\xi}_3) \\ &\quad + \kappa(\xi_3' + \bar{\xi}_3') + B \end{aligned} \tag{182}$$

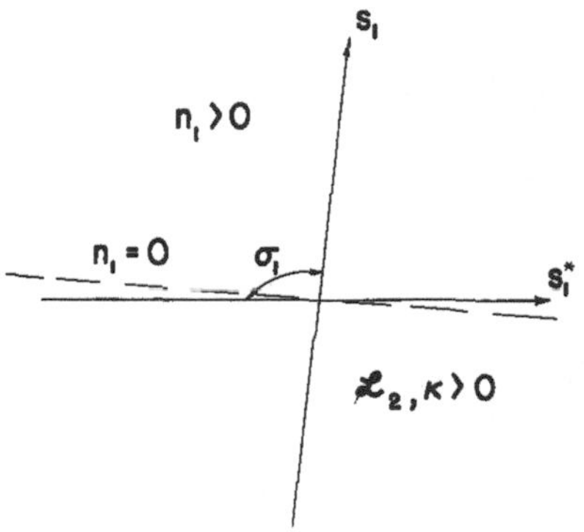

FIG. 86a

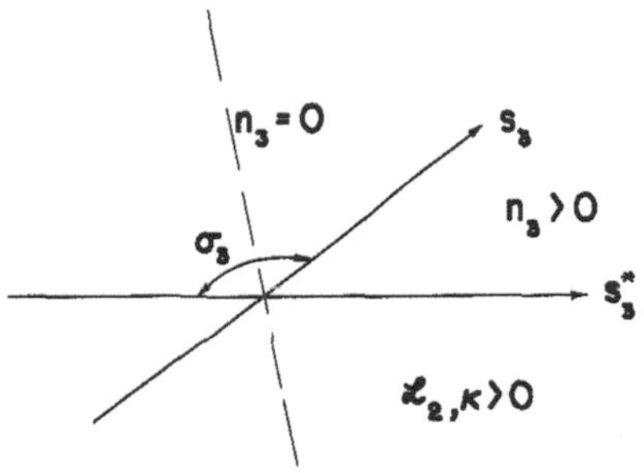

FIG. 86b

For ease in writing we introduce

$$\left.\begin{aligned} s_k &= \xi_k + \bar{\xi}_k \\ s_k^* &= \frac{\xi_k' + \bar{\xi}_k'}{|\lambda_k|} \end{aligned}\right\} \tag{183}$$

which leads to

$$\left.\begin{aligned} \theta &= \sum_{k=1,3} s_k + B \\ \theta' &= \sum_{k=1,3} s_k^* |\lambda_k| \\ F = \theta + \kappa\theta' &= \sum_{k=1,3}[s_k + \kappa |\lambda_k| s_k^*] + B \end{aligned}\right\} \tag{184}$$

All the necessary formulas have been given. We are now in a position to explain the use of "partial" phase planes. Two phase planes are employed, one with the coordinates s_1 and s_1^* forming the angle σ_1 (see equation (175d)) with each other, and a second with the coordinate s_3 and s_3^* forming the angle σ_3 with each other. The locus of s_k and s_k^*, as given by equations (183) and (175a, c), is a spiral in the corresponding phase plane with the center (0,0). Knowing the initial values of s_k and s_k^*, we may construct the "partial phase curves."

The coordinates s_k and s_k^* are both discontinuous at the switch points.

The jumps are given by equations (180) and (181). The locations of those points in the different partial phase planes which correspond to the switch point in the original phase space (where the crossing of $F = 0$ occurs) have yet to be found. Looking at equation (184) for the control function, we observe

$$F = s_1 + \kappa|\lambda_1|s_1^* + s_3 + \kappa|\lambda_3|s_3^* + B \tag{185}$$

$F = 0$ means that

$$s_1 + \kappa|\lambda_1|s_1^* + B = -(s_3 + \kappa|\lambda_3|s_3^*) \tag{186}$$

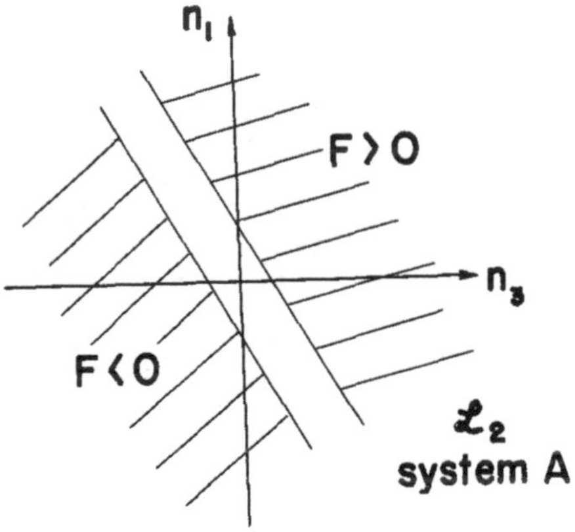

FIG. 87a

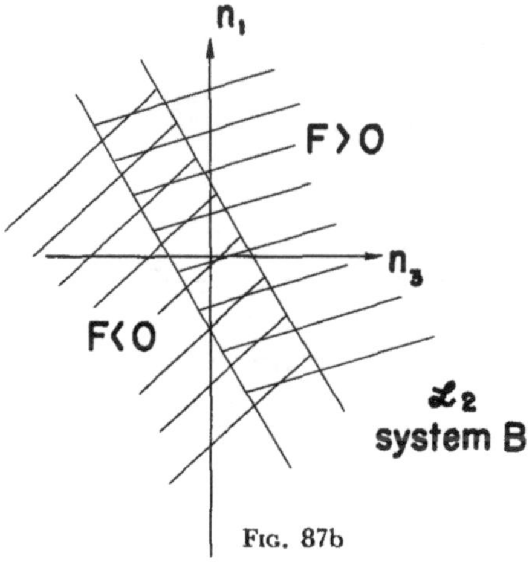

FIG. 87b

The terms

$$F_k = s_k + \kappa|\lambda_k|s_k^* = 0 \tag{187}$$

represent straight lines in the corresponding phase planes as seen in Figs. 86a and 86b. The normal distance n_k of a spiral point from $F_k = 0$ is given by

$$n_k = \frac{\sqrt{1 - D_k^2}}{\sqrt{1 - 2\kappa|\lambda_k|D_k + \kappa^2|\lambda_k|^2}}[s_{k_{sp}} + \kappa|\lambda_k|s_{k_{sp}}^*]$$

where $-D_k = \cos\sigma_k$.

Thus $F = 0$ corresponds to

$$\frac{n_1\sqrt{1 - 2\kappa|\lambda_1|D_1 + \kappa^2|\lambda_1|^2}}{\sqrt{1 - D_1^2}} + B = \frac{-n_3\sqrt{1 - 2\kappa|\lambda_3|D_3 + \kappa^2|\lambda_3|^2}}{\sqrt{1 - D_3^2}} \tag{188}$$

An auxiliary diagram, such as shown in Figs. 87a and 87b, becomes necessary to determine $F = 0$ in the partial phase planes. Note that the diagram for system A shows a gap between the regions $F > 0$ and $F < 0$. For system B there exists an overlapping of these regions, an immediate consequence of which is that rest of control may occur only in system B. Rest of control requires that spirals around the origin of the s_1, s_1^* and s_3, s_3^* plane reach the origin of the respective plane for $t \to \infty$. Hence very small values n_1 and n_3 will appear, tending towards zero with increasing time. Only in the auxiliary diagram for system B can very small values of n_1 and n_3 exist simultaneously for either $F < 0$ or $F > 0$.

In the next section an example will be studied, and the construction outlined above will be illustrated in detail.

8.4. Examples. By way of illustration we will show the motion of a particular missile under discontinuous position control. Most of the examples are taken from an earlier work[1] in which the results were obtained by computation instead of by a partial phase plane construction as used here.

The physical constants of the missile are given in Table VIII, and the initial values of the variables for each example are given in the corresponding figure. All of the examples will refer to aileron control (system L_2).

We have already given several results obtained from one example (see Sec. 8.2 and Fig. 85). Now we will study a very similar example.

Before proceeding to the actual computations and constructions, it will help to write the solution in a slightly different form. Let us rearrange the right sides of equations (172) by stressing the position of the initial conditions and write:

$$\theta - B = \sum_{k=1}^{4} e^{\lambda_k \tau} \left[\left(\frac{v_i}{V_0} - C \right) \mathfrak{C}_{1k} + (\alpha_i - A)\mathfrak{C}_{2k} + (\theta_i - B)\mathfrak{C}_{3k} + \theta_i' \mathfrak{C}_{4k} \right] \tag{189}$$

where the $\mathfrak{C}_{1k} \ldots \mathfrak{C}_{4k}$ are constants which may be computed with the aid of the subdeterminants of equations (172). Their numerical values are given in Table VIII. Similarly, the other solutions may be written as

$$\alpha - A = \sum_{k=1}^{4} A_k e^{\lambda_k \tau} \left[\left(\frac{v_i}{V_0} - C \right) \mathfrak{C}_{1k} + (\alpha_i - A)\mathfrak{C}_{2k} + (\theta_i - B)\mathfrak{C}_{3k} + \theta_i' \mathfrak{C}_{4k} \right] \tag{190}$$

$$\frac{v}{V_0} - C = \sum_{k=1}^{4} C_k e^{\lambda_k \tau} \left[\left(\frac{v_i}{V_0} - C \right) \mathfrak{C}_{1k} + (\alpha_i - A)\mathfrak{C}_{2k} + (\theta_i - B)\mathfrak{C}_{3k} + \theta_i' \mathfrak{C}_{4k} \right] \tag{191}$$

[1] I. Flügge-Lotz and H. Meissinger, Über Bewegungen eines Schwingers unter dem Einfluss von Schwarz-Weiss-Steuerungen, IV. Bewegungen eines Schwingers von drei Freiheitsgraden, untersucht am Beispiel der Flugzeuglängsschwingung; Steuerung mit Stellungszuordnung. *Zentrale für wissenschaftliches Berichtswessen der Luftfahrtforschung des Generalluftzeugmeisters (ZWB)*, Untersuchungen und Mitteilungen Nr. 1329, Berlin, August 10, 1944.

For computing the initial values of s_k and s_k^* we must know the initial magnitude of F. This is found from the equation $F_i = \theta_i + \kappa\theta_i'$.

Now let us return to the example. The complete results of the construction are given in Fig. 88. Using the initial values θ_i and θ_i' given in the figure, we find that $F_i < 0$. Choosing control system B, the initial values

$$s_{1i} = -.63, \qquad s_{3i} = +.055$$

$$s_{1i}^* = -.69, \qquad s_{3i}^* = 0$$

are easily computed. After locating the initial points in the partial phase planes, we construct the appropriate spirals about the origins of the coordinate axes, as shown in Fig. 88.

The points where the control function switches in the phase planes are found as described on p. 136 (see Fig. 87b). Here again the construction is shown in Fig. 88. The magnitude of the jumps is given by equations (180) and (181), and they are found to be

$$\Delta s_1 = -.0507 \quad (\text{sgn}\, F)$$

$$\Delta s_3 = +.00470 \,(\text{sgn}\, F)$$

$$\Delta s_1^* = +.304 \quad (\text{sgn}\, F)$$

$$\Delta s_3^* = -.00713 \,(\text{sgn}\, F)$$

As already mentioned on p. 134 these values are missile constants and do not change for different examples, provided that the maximum aileron angle η_0 is held constant.

The construction shows that in most cases the end of the interval will be determined by the low frequency partial motion. The influence of the high frequency motion is important only behind the switch points. The time dependent diagrams of $\theta(\tau)$ and $\theta'(\tau)$ are easily obtained from the partial phase planes in the same manner as in the case of a body with one degree of freedom. The final state of motion is a periodic one, practically attained after a few intervals.

The remaining examples will not be used to illustrate the construction particularly, although much may be gained from studying the figures, but they will show characteristic partial phase curves and their approach to an asymptotic state. In order to simplify the illustrations, only certain intervals of each motion are shown. The actual start of the motion (which is rarely a switch point) is omitted. Each example shows the two partial phase planes, the auxiliary diagram, and a time dependent diagram of the variables. The functions $\theta(\tau)$ and $\theta'(\tau)$ are readily obtained from the partial phase planes. If the functions $\alpha(\tau)$ and $v/V_0(\tau)$ are also desired (for computing the

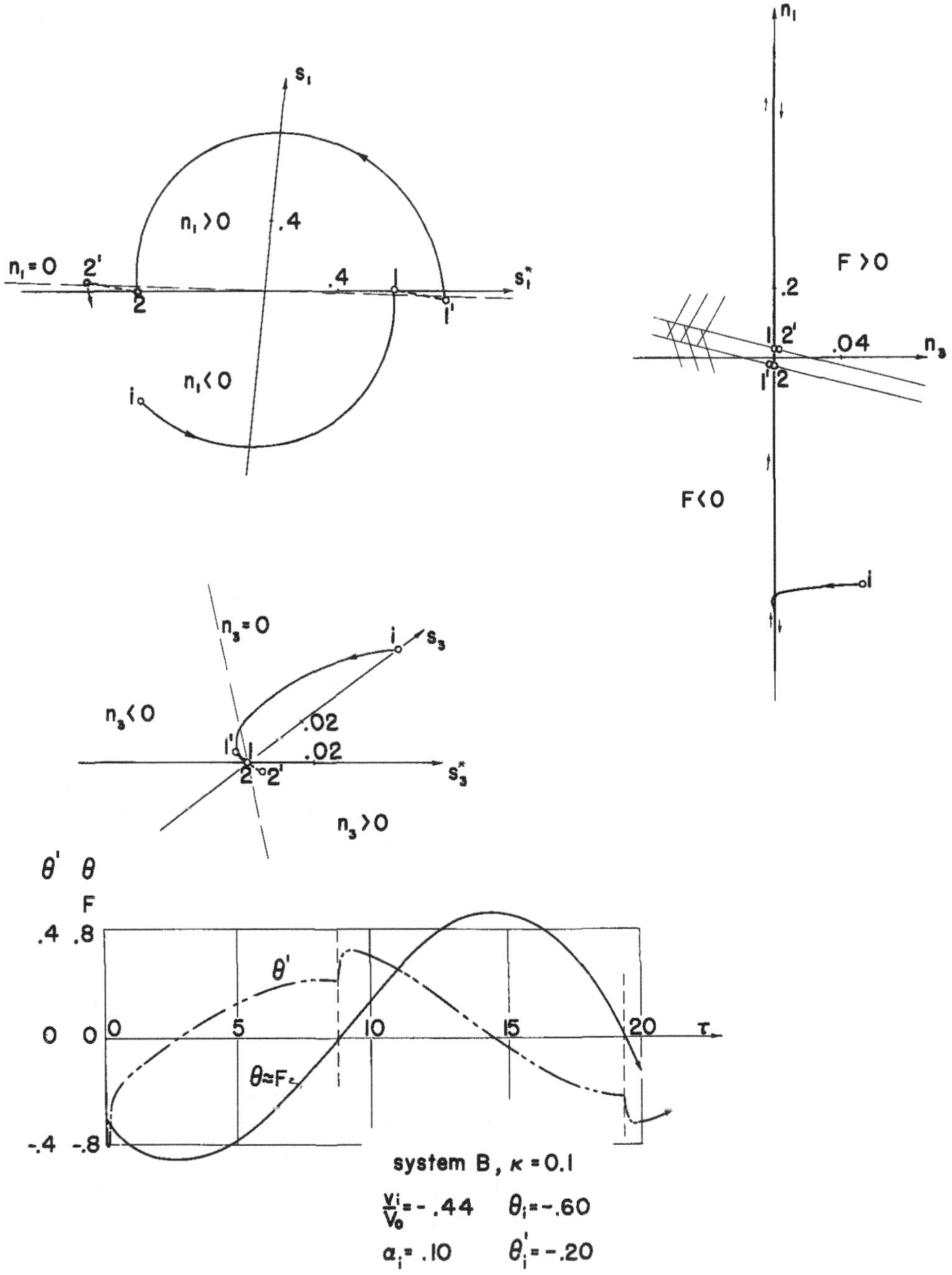

n₁
F > 0
.2
1 2'
.04
n₃
1' 2
F < 0
i
s₁
n₁ > 0
.4
n₁ = 0
2'
2
.4
1
s₁*
1'
n₁ < 0
i
n₃ = 0
i
s₃
n₃ < 0
.02
1'
1
.02
2
2'
s₃*
n₃ > 0
θ' θ
F
.4 .8
0 0
-.4 -.8
θ'
θ≈F
0
5
10
15
20
τ
system B, κ = 0.1
vi/V₀ = -.44 θi = -.60
αi = .10 θ'i = -.20

FIG. 88

path of the missile), either one of two methods for obtaining them may be employed. These are outlined as follows:

(a) The system of equations (176) may be regarded as a system for the unknowns $C_k\xi_k$ or for ξ_k. If we do consider $C_k\xi_k$ as the unknowns, we may choose two other partial phase planes having the coordinates

$$\tilde{s}_1 = C_1\xi_1 + \overline{C_1\xi_1}, \qquad \tilde{s}_1^* = \frac{C_1\xi_1' + \overline{C_1\xi_1'}}{|\lambda_1|}$$

and

$$\tilde{s}_3 = C_3\xi_3 + \overline{C_3\xi_3}, \qquad \tilde{s}_1^* = \frac{C_3\xi_3' + \overline{C_3\xi_3'}}{|\lambda_3|}$$

The initial values of $\tilde{s}_k$ and $\tilde{s}_k^*$ are easily computed, and the angles σ_k of the coordinate axes are known already. The jumps of $\tilde{s}_k$ and $\tilde{s}_k^*$ may be computed analogously to those of s_k and s_k^*. Thus the construction of the partial phase curves may begin. The lengths of the intervals are already known from the earlier construction in the s_k, s_k^* planes. The function $v(\tau)/V_0$ would be easily obtained from this construction, and $\alpha(\tau)$ would be obtained by using the first equation of system L_2 (168a). Compare this with equation (165) for p_{11}, p_{12}, and p_{13}.

(b) Because the lengths of the intervals are known, $\alpha(\tau)$ and $v(\tau)/V_0$ may be computed by using equations (190) and (191), always using the θ, θ', α, and v/V_0 values at the end of one interval for the start of the next one. (see p. 128/9.)

Method (a) gives a better mechanical view of the problem, but it may appear to be more tedious than method (b) to those who are not familiar with graphical constructions.

The second example (Fig. 89) leads to a periodic motion also, which is practically reached at the point 3′. Although there are large changes in the variables, it can be seen that here too the high frequency oscillation is significant near the switch points only. The essential difference between this example and the preceding one is the shorter period of the asymptotic periodic motion. In the preceding example one may attempt to give an approximation to the asymptotic behavior of θ and θ' by a simple sine wave. With the exception of the neighborhood of the switch point, the deviation would not be too large. In the present example such an approximation would be much less accurate.

Fig. 90 presents an example in which the high-frequency oscillation plays a predominant role because the first switch point shown is given by the high-frequency motion. This result shows the importance of including the high-frequency oscillation in the analysis. After the first switch point the amplitudes increase appreciably. Nevertheless the final state is a periodic motion. (In the study of single degree of freedom systems we met motions which approached a periodic state by increasing amplitudes as well as by decreasing ones).

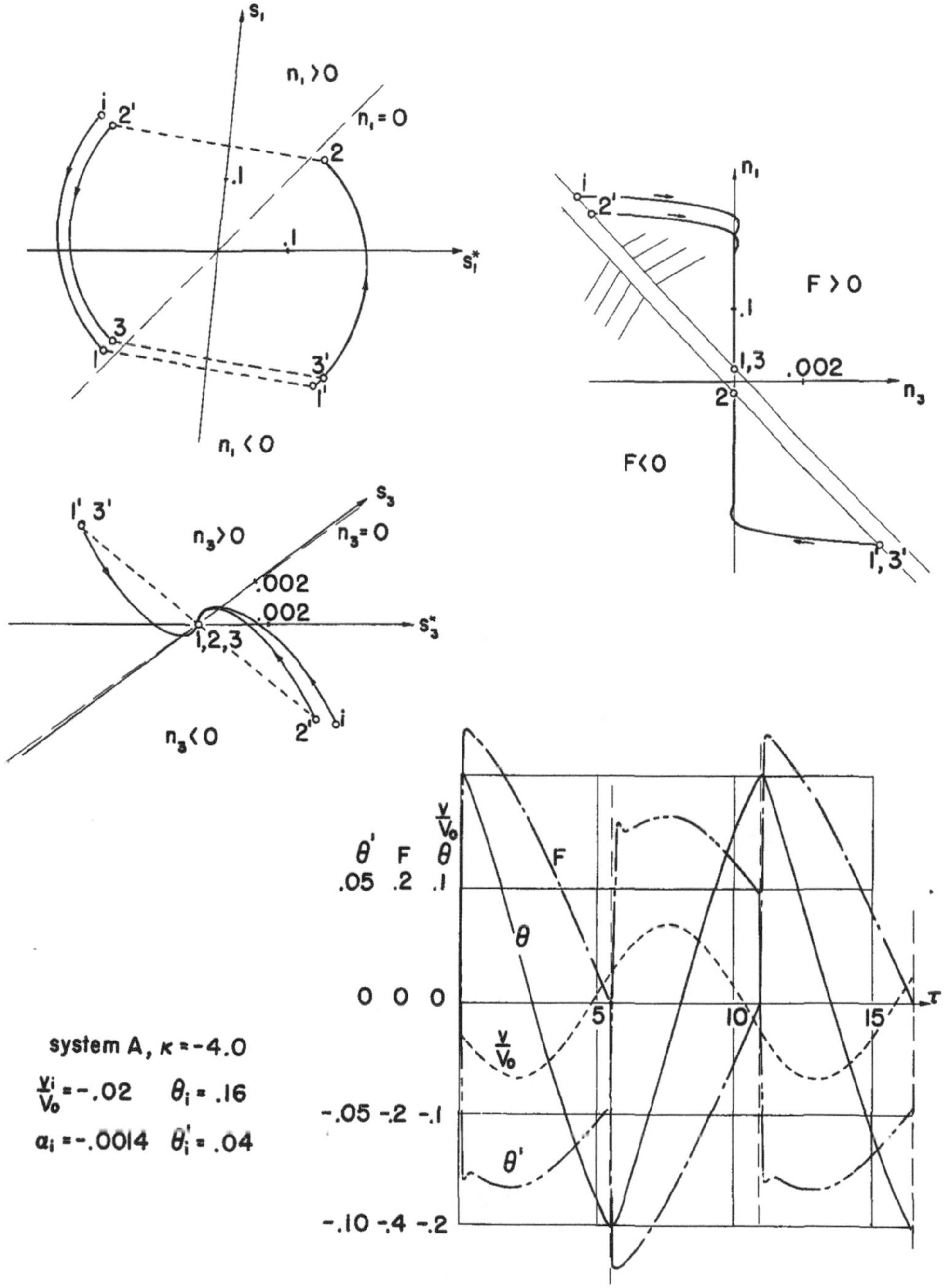
s_1
$n_1 > 0$
$n_1 = 0$
$n_1 < 0$
s_1^*
n_1
F > 0
F < 0
.002
n_3
s_3
$n_3 > 0$
$n_3 = 0$
$n_3 < 0$
s_3^*
1,2,3
$\frac{v}{V_0}$
θ' F θ
.05 .2 .1
0 0 0
-.05 -.2 -.1
-.10 -.4 -.2
5
10
15
τ
system A, $\kappa = -4.0$
$\frac{v_i}{V_0} = -.02$ $\theta_i = .16$
$\alpha_i = -.0014$ $\theta_i' = .04$

FIG. 89

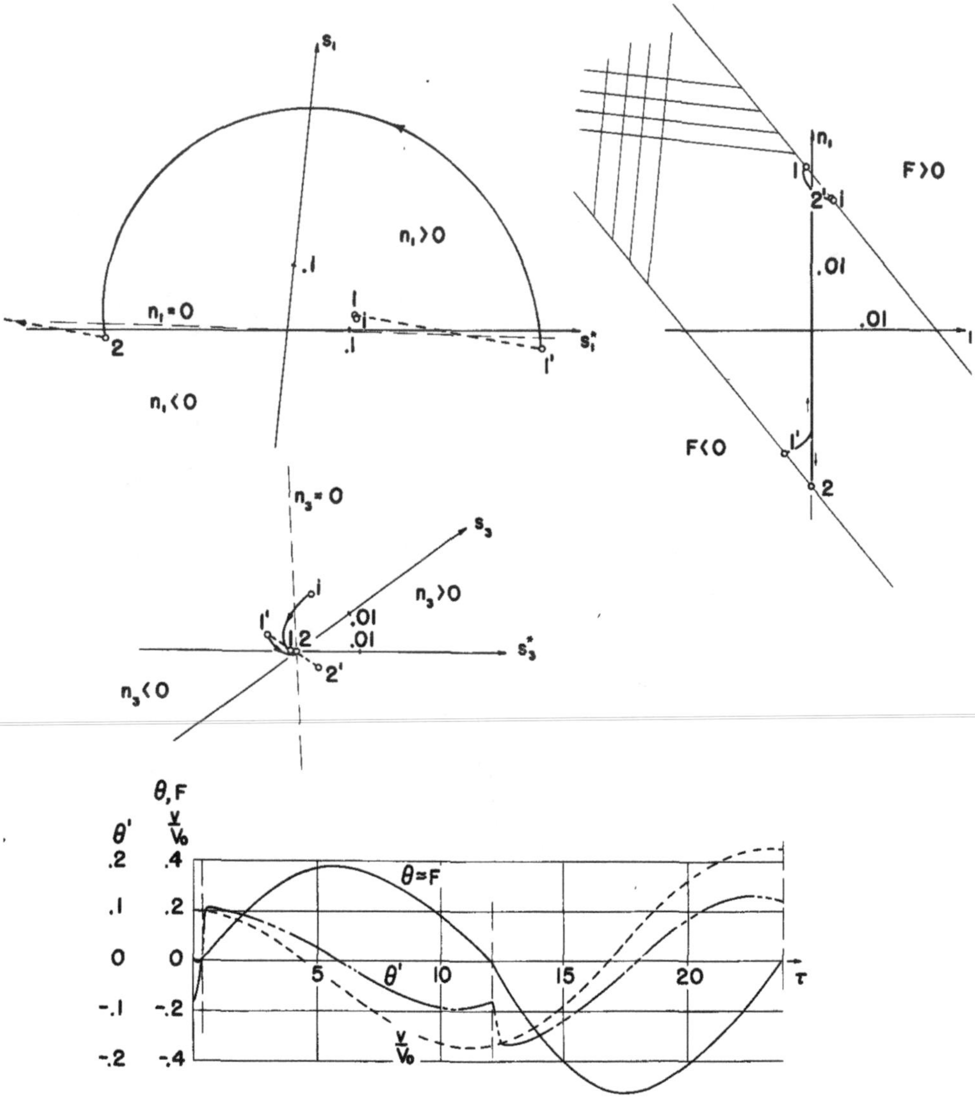

system B, $\kappa = 0.1$

$\frac{v_i}{V_0} = .20 \quad \theta_i = .008$

$\alpha_i = .018 \quad \theta_i' = -.08$

FIG. 90

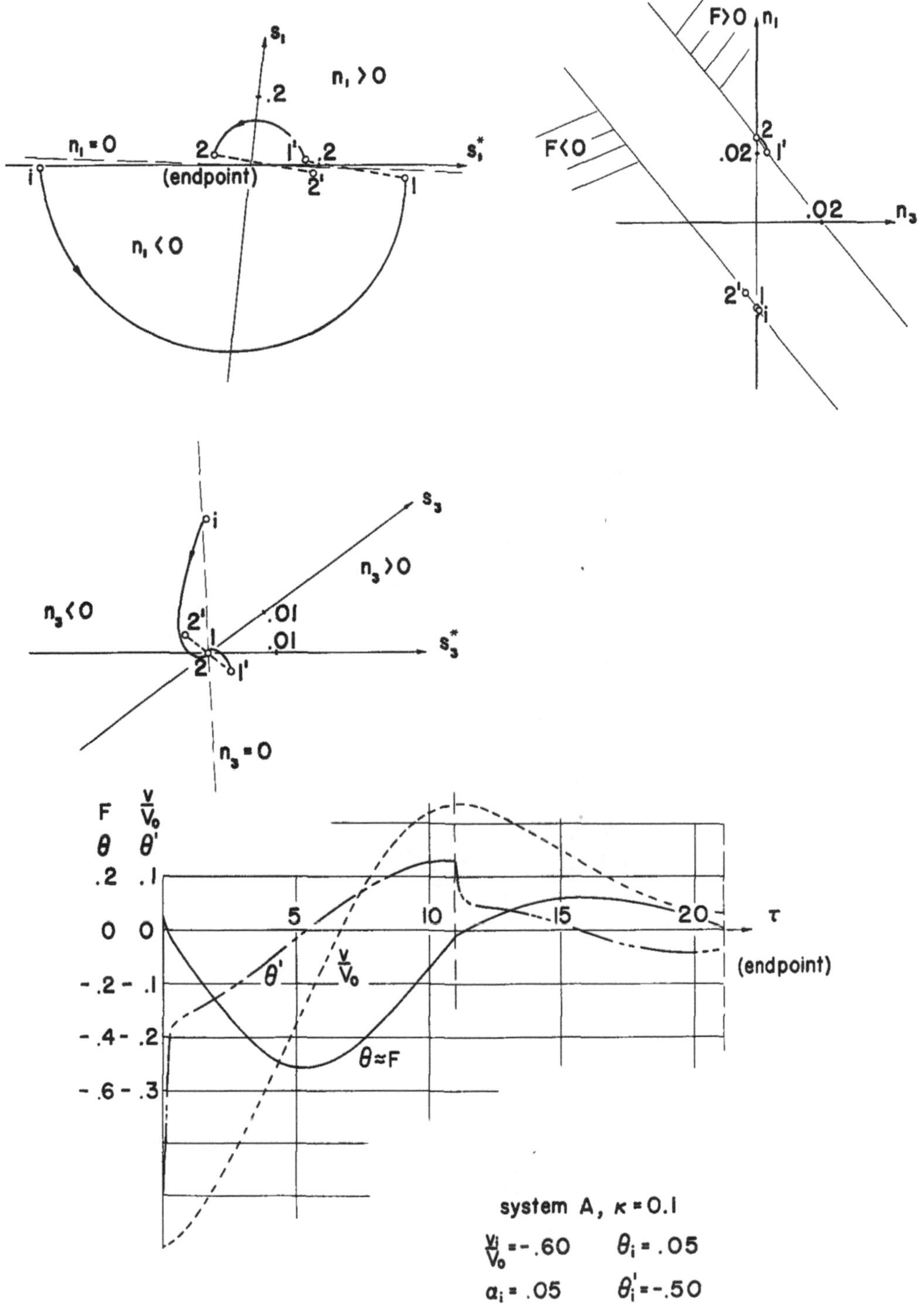

s1
n1 > 0
.2
n1 = 0
2
1'
.2
s1*
i
(endpoint)
2'
1
n1 < 0
F>0
n1
F<0
2
.02
1'
.02
n3
2'
1
i
s3
i
n3 > 0
n3 < 0
.01
2'
1
.01
s3*
2
1'
n3 = 0
F
v/V0
θ
θ'
.2
.1
5
10
15
20
τ
0
0
-.2
-.1
θ'
v/V0
(endpoint)
-.4
-.2
θ≈F
-.6
-.3
system A, κ = 0.1
vi/V0 = -.60
θi = .05
αi = .05
θ'i = -.50

FIG. 91

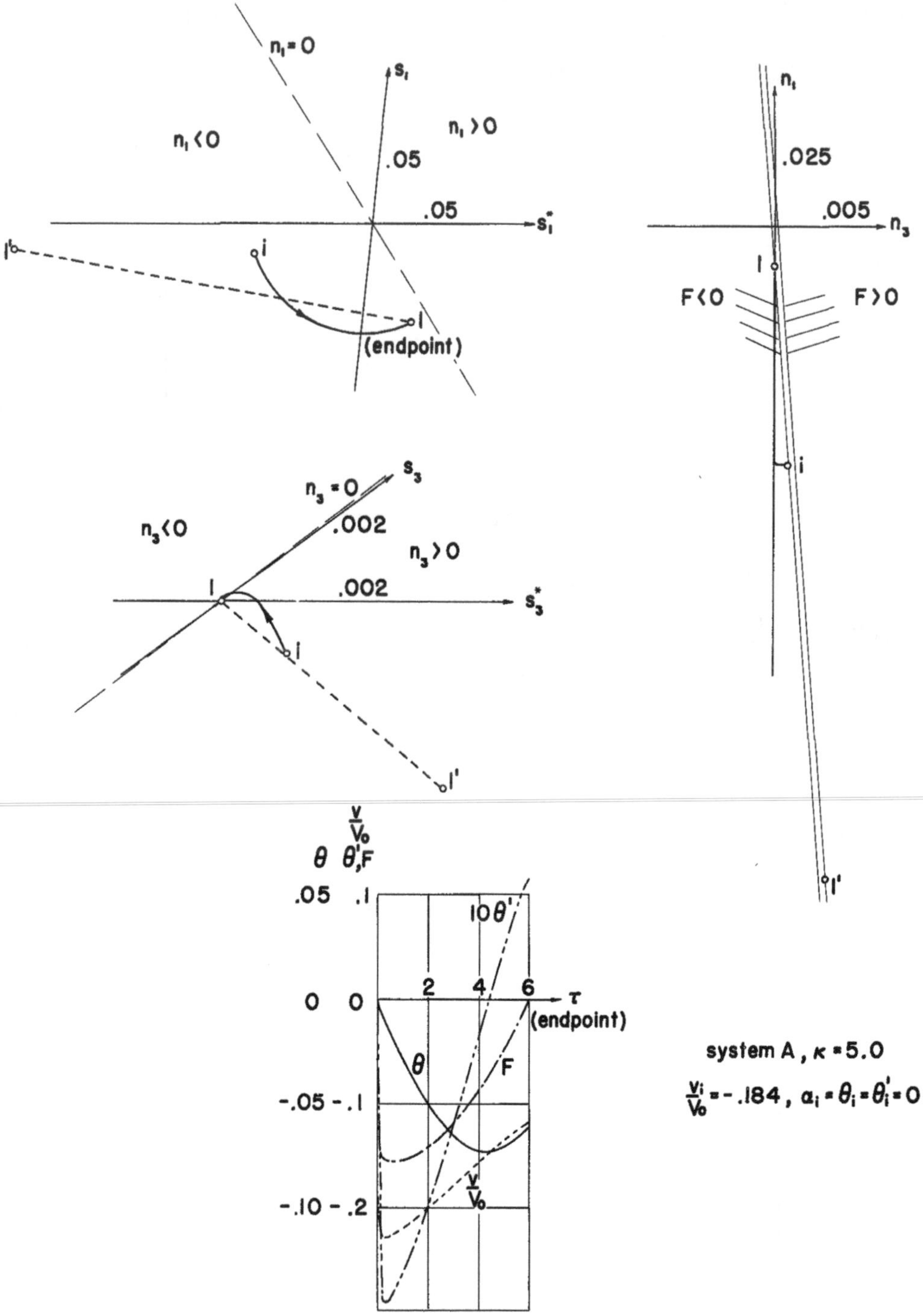

n₁ = 0
s₁
n₁ < 0
n₁ > 0
.05
.05
s₁*
l'
i
l
(endpoint)
n₁
.025
.005
n₃
l
F < 0
F > 0
i
l'
s₃
n₃ = 0
.002
n₃ < 0
n₃ > 0
l
.002
s₃*
i
l'
v/V₀
θ θ', F
.05 .1
10θ'
2 4 6 τ
0 0
(endpoint)
θ
F
-.05 -.1
v/V₀
-.10 -.2
system A, κ = 5.0
vᵢ/V₀ = -.184, αᵢ = θᵢ = θᵢ' = 0

Fig. 92

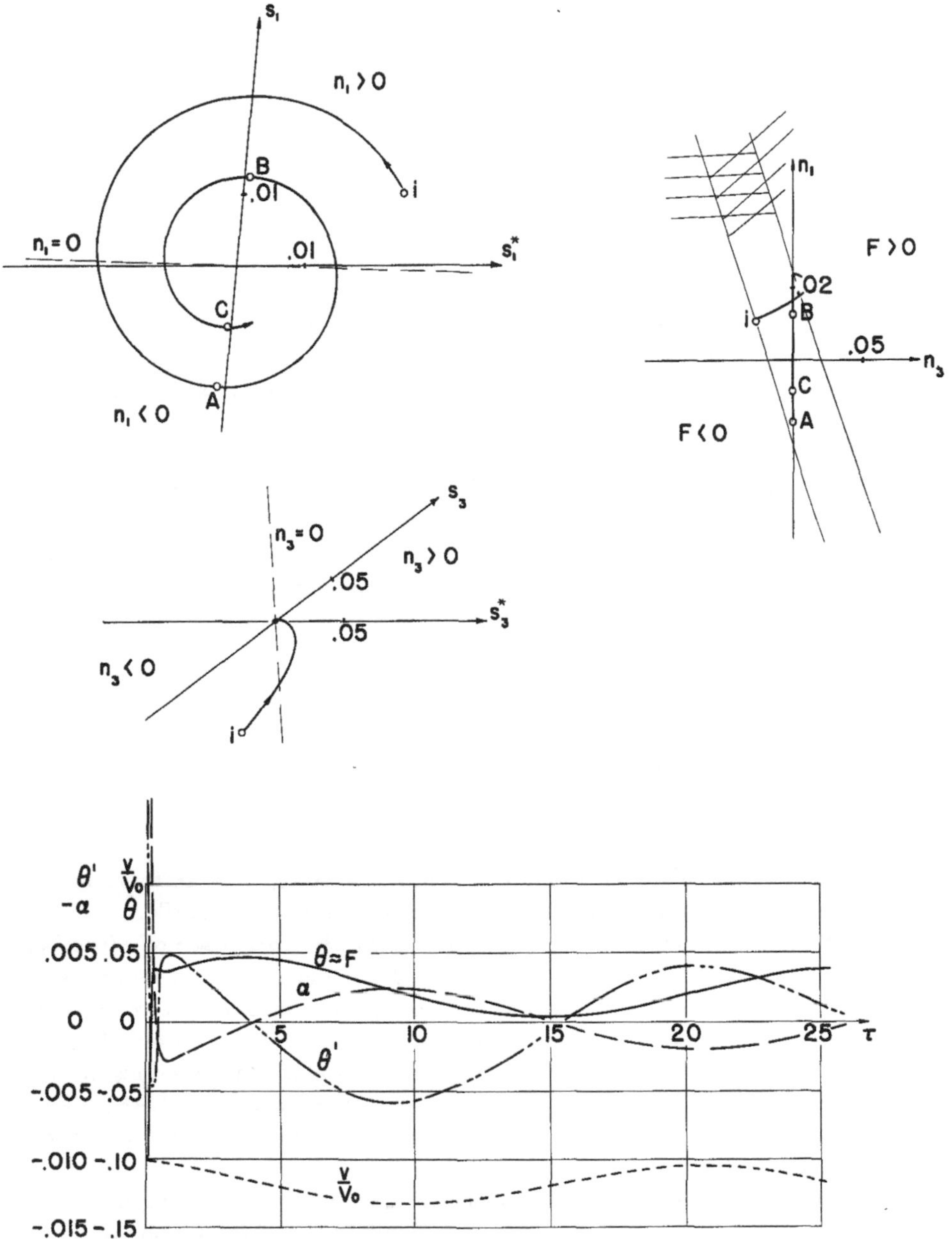

system B, $\kappa = 0.1$

$\frac{v_i}{V_0} = -.10 \quad \theta_i = -.10$

$\alpha_i = -.18 \quad \theta_i' = 1.0$

FIG. 93

Figs. 91 and 92 present two examples which go to end points. In Fig. 91 the coefficient $\kappa = 0.1$, and in Fig. 92 $\kappa = 5.0$. In Fig. 91 the motion proceeds quite regularly at first. After arriving at point 2 the necessary jump is put in and point 2′ is obtained. But point 2′ has a negative value of n_1, and a continuation of the spiral in the s_1, s_1^* plane would require smaller negative values of n_1 with rapidly decreasing $|n_3|$. This would lead to that region of the auxiliary diagram for which the motion is undefined. Thus it is clear that the point 2 is an end point for the phase curves, so long as there is no delay between the switching and $F = 0$.

In Fig. 92 the end point occurs at the end of the first interval. If at point 1′ spirals are tentatively started in the partial phase planes, we would see that $|n_1|$ would increase and $|n_3|$ would decrease. The unavoidable consequence would be that the motion would lead into the gap between $F > 0$ and $F < 0$.

An example showing a motion which goes immediately to rest of control is given in Fig. 93. System B, as already mentioned, permits "control at rest," thus $n_1 \to 0$ and $n_3 \to 0$ simultaneously. For the convenience of the reader some special points in the partial phase planes are noted in the n_1, n_3 plane. The asymptotic state of this motion is (because $F > 0$)

$$\theta_\infty = B = \frac{c_{10}}{b_{10}c_{20} - c_{10}b_{20}} N_2\eta_0$$

$$\left(\frac{v}{V_0}\right)_\infty = -\frac{b_{10}}{b_{10}c_{20} - c_{10}b_{20}} N_2\eta_0$$

$$\alpha_\infty = 0$$

It is clear that this is an undesired asymptotic state, because there results an infinitely large deviation from the original path of the missile.

8.5. The occurrence of end points and the influence of time lag on the controlled motion. In our study of the controlled motion of a body with one degree of freedom we recognized the importance of those regions in the phase plane where with ideal control the motions would become undefined. We learned there might be favorable imperfections in the control mechanism which would make part of those regions the best working regions. These ideas will guide the investigation for the body with three degrees of freedom.

Figs. 94 and 95 show the influence of time lag on two examples of the motion of a body with three degrees of freedom. Justified by our previous experience, we assume that time lag does not greatly effect the general features of a phase curve as long as the end point region is not entered. In these examples the time lag was first taken into account at that time when the motion would become undefined for ideal control. Fig. 94 shows a motion with $\kappa = 0.1$ and a time lag of $\tau_r = 0.25$ (refer to Fig. 91 for the motion with ideal control). Fig. 95 shows a motion with $\kappa = 5.0$ and two

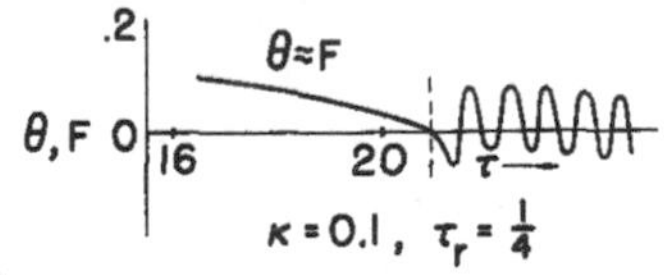
.2
θ≈F
θ,F 0
16
20
τ
κ = 0.1 , $\tau_r = \frac{1}{4}$

FIG. 94

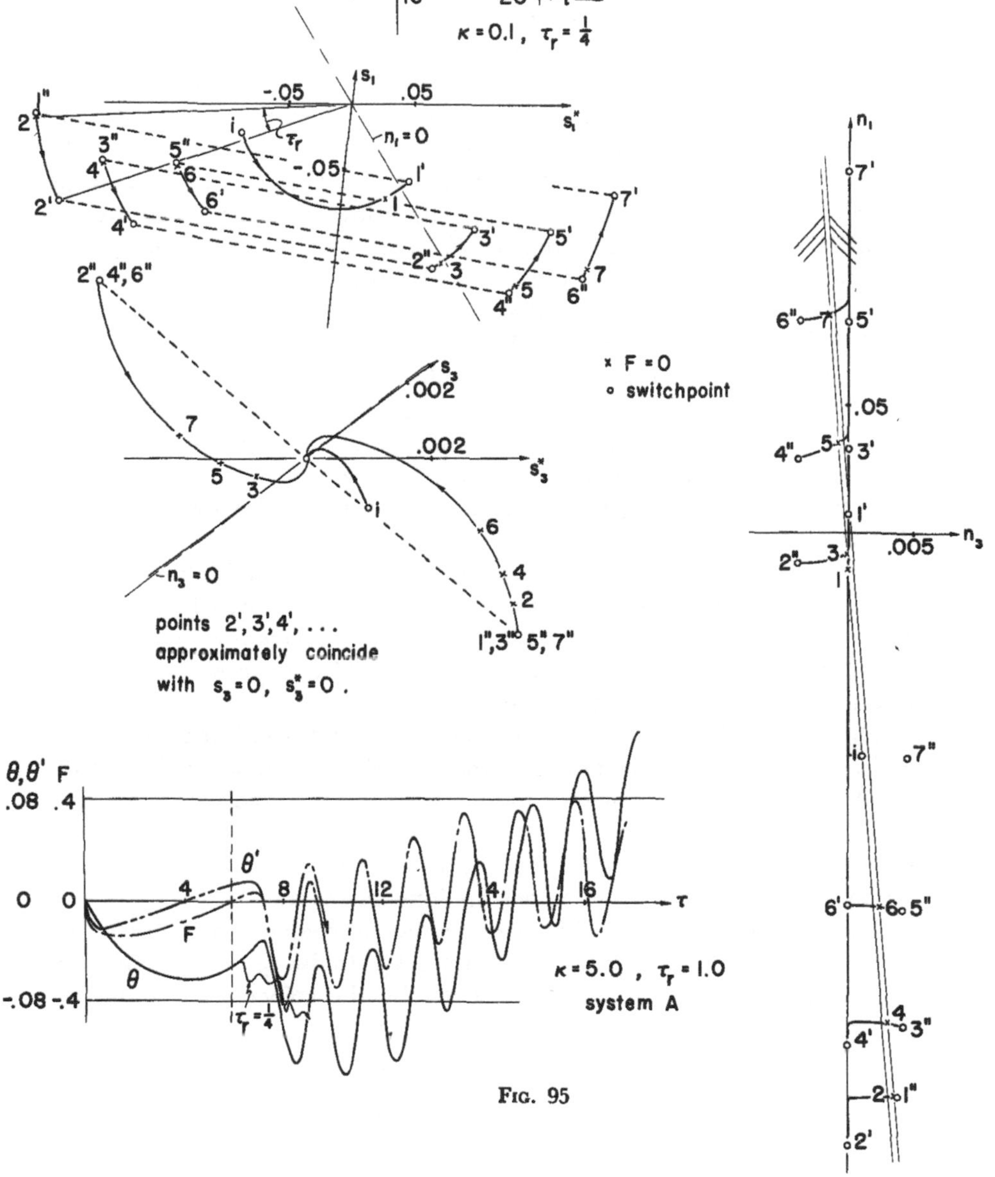
s_1
-.05
.05
s_1^*
$n_1 = 0$
τ_r
× F = 0
○ switchpoint
s_3
.002
.002
s_3^*
$n_3 = 0$
points 2', 3', 4', . . .
approximately coincide
with $s_3 = 0$, $s_3^* = 0$.
1", 3", 5", 7"
2", 4", 6"
n_1
.05
.005
n_3
θ,θ' F
.08 .4
0 0
-.08 -.4
θ'
F
θ
$\tau_r = \frac{1}{4}$
τ
κ = 5.0 , τ_r = 1.0
system A

FIG. 95

different sizes of time lag, $\tau_r = 1.0$ and $\tau_r = 0.25$ (for the motion with ideal control, refer to Fig. 92). In Fig. 95 the partial phase planes are shown at least for a short length of time following the end point, in order to demonstrate the construction involved. Then the curves of θ and F versus time are given. The θ' curve is not given completely because it follows the curve of F closely. In Fig. 94 only the curve of $\theta(\simeq F)$ versus time is shown.

In both of these examples the "after-end point" motion has high frequency. The amplitude of this high-frequency oscillation depends essentially upon the size of the time lag. However, one essential difference between the two examples should be noted. In Fig. 94 ($\kappa = 0.1$) the average value of the control function F is approximately zero. In Fig. 95 ($\kappa = 5.0$) the average value of F is described by a cosine line with a low frequency. The average behavior of θ corresponds to that of F. Therefore the "after-end point" motion is desirable in Fig. 94 but not in Fig. 95.

The fact that this different behavior is essentially due to the size of the coefficient κ should not be difficult to understand. For very small values of κ the functions θ and F are nearly in phase (see Fig. 91). This is not the case for large κ (see Fig. 92). Then for small κ the zero points of θ nearly coincide with the zero points of F. This means that the high-frequency motion starts with a value of θ which is nearly zero. Now because θ and F are very nearly in phase, θ can never attain a strong deviation from zero if τ_r is small.

In conclusion, as long as a strong damping of undesirable disturbances is sought, a small $|\kappa|$ should be taken. Yet it would not be advisable to go to the limit to suggest $\kappa = 0$. If $\kappa = 0$ we would have $F = 0$ and $\theta = 0$ simultaneously at the switch points, but we could not expect an end point (see the section on continuity qualities).

The example in Fig. 94 shows that the "after-end point" motion may be a desired state of motion provided that the time lag is so small that the amplitude of the high-frequency motion is small compared with the amplitude of θ in the preceding intervals. Hence it is important to study the occurrence of end points in the phase space for ideal control and the average motion after those end points under the influence of time lag.

The motion of the body with three degrees of freedom may be described in a phase space with the axes θ, θ', α, and v/V_0. In this space the boundaries of the end point region would be given by the coincidence of the two expressions

$$\left.\begin{aligned} F &= \theta + \kappa\theta' = 0 \\ F' &= \theta' + \kappa\theta'' = 0 \end{aligned}\right\} \qquad (192)$$

Eliminating κ we could find those portions of the space where switch points that are end points occur. However, because our imagination is limited to three-dimensional space, a formal description by mathematical equations would be required. On the other hand, we are interested only in those regions where, under the influence of imperfections, high-frequency motions that oscillate around an average phase curve leading back to $\theta = \theta' = \alpha = v/V_0 = 0$ occur. For small $|\kappa|$ this average motion may be found by assuming $F = 0$.

The average motion for the case of control by ailerons (L_2) is described by the following set of equations:

$$\left.\begin{aligned} &p_{11}\alpha + p_{12}\theta + p_{13}\frac{v}{V_0} = 0 \\ &p_{21}\alpha + p_{22}\theta + p_{23}\frac{v}{V_0} = -N_2\eta \\ &p_{31}\alpha + p_{32}\theta = 0 \\ &F = \theta + \kappa\theta' = 0 \end{aligned}\right\} \tag{193}$$

The strong connection between the angle of the aileron and the control function that we had earlier becomes meaningless if we consider the average motion. Thus from the fact that three equations from (193) will determine $\alpha_a(\tau)$, $\theta_a(\tau)$, and $v_a(\tau)/V_0$, we will have

$$\left.\begin{aligned} &p_{11}\alpha_a + p_{12}\theta_a + p_{13}\frac{v_a}{V_0} = 0 \\ &p_{31}\alpha_a + p_{32}\theta_a = 0 \\ &\theta_a + \kappa\theta_a' = 0 \end{aligned}\right\} \tag{194}$$

Now we may compute a continuous function $\eta(\tau)$ from the second equation of system (193)

$$p_{21}\alpha_a + p_{22}\theta_a + p_{23}\frac{v_a}{V_0} = -N_2\eta_a \tag{195}$$

Thereby we obtain an aileron angle η_a which would aid in creating a continuous missile motion corresponding to the average motion (that exists only in our imagination).

The characteristic determinant of equations (194) gives the roots needed for establishing the solutions:

$$\begin{vmatrix} a_{10} & b_{10} & c_{10} + c_{11}\lambda_a \\ a_{30} + a_{31}\lambda_a & b_{31}\lambda_a + b_{32}\lambda_a^2 & 0 \\ 0 & 1 + \kappa\lambda_a & 0 \end{vmatrix} = 0 \tag{196}$$

or

$$(1 + \kappa\lambda_a)\,(a_{30} + a_{31}\lambda_a)\,(c_{10} + c_{11}\lambda_a) = 0 \tag{197}$$

The existence of three roots is indicated. The three roots are

$$\lambda_{a_1} = -\frac{a_{30}}{a_{31}} = -\frac{\bar{\mu}C'_{M_0}}{C_{Mq} \cdot k_1} \tag{198}$$

(which is less than zero, if the missile is stable without control)

$$\lambda_{a_2} = -\frac{c_{10}}{c_{11}} = -2C^*_{D0} < 0 \tag{199}$$

and

$$\lambda_{a_3} = -\frac{1}{\kappa} \tag{200}$$

which is negative if κ is chosen positive.

We may adapt the average motion to the initial conditions, given approximately by the values of θ, α, and v/V_0 at the end point of the motion under ideal control. The result of the preceding computation is interesting, for it shows that it is impossible to have a favorable average motion of an unstable missile. In this event λ_{a_1} would be positive.[1]

In the case of elevator control the average motion for small $|\kappa|$ is described by

$$\left.\begin{aligned} p_{11}\alpha + p_{12}\theta + p_{13}\frac{v}{V_0} &= 0 \\ p_{21}\alpha + p_{22}\theta + p_{23}\frac{v}{V_0} &= 0 \\ p_{31}\alpha + p_{32}\theta &= N_3\eta_h \\ F = \theta + \kappa\theta' &= 0 \end{aligned}\right\} \tag{201}$$

The values of $\alpha_a(\tau)$, $\theta_a(\tau)$, and $v_a(\tau)/V_0$ will be determined by the three equations

$$\left.\begin{aligned} p_{11}\alpha_a + p_{12}\theta_a + p_{13}\frac{v_a}{V_0} &= 0 \\ p_{21}\alpha_a + p_{22}\theta_a + p_{23}\frac{v_a}{V_0} &= 0 \\ \theta_a + \kappa\theta'_a &= 0 \end{aligned}\right\} \tag{202}$$

The equation

$$p_{31}\alpha_a + p_{32}\theta_a = N_3\eta_{h_a} \tag{203}$$

will be used to compute a continuous function η_{h_a}, the elevator angle which

[1] In case that the effect of time lag between the creation of downwash by the wing and its action on the tail is neglected, the term $a_{31}\lambda_a$ would not appear in equation (196) and equation (197) would be reduced to $(c_{10} + c_{11}\lambda_a)(1 + \kappa\lambda_a) = 0$. Thus we would not recognize the restriction of needing a basically stable missile.

might produce a continuous motion of the missile equivalent to the average motion. The roots needed for the solution of equations (202) are given by

$$\begin{vmatrix} a_{10} & b_{10} & c_{10} + c_{11}\lambda_a \\ a_{20} + a_{21}\lambda_a & b_{20} + b_{21}\lambda_a & c_{20} \\ 0 & 1 + \kappa\lambda_a & 0 \end{vmatrix} = 0 \tag{204}$$

or $$(1 + \kappa\lambda_a)\,[(a_{10}c_{20} - a_{20}c_{10}) - \lambda_a(a_{21}c_{10} + c_{11}a_{20}) - a_{21}c_{11}\lambda_a^2] = 0 \tag{205}$$

Equation (205) shows at once that for this control as well $\kappa > 0$ is an unavoidable condition for stability. It may be mentioned here that the other two roots are negative and that the roots are all independent of C'_{M_0}.

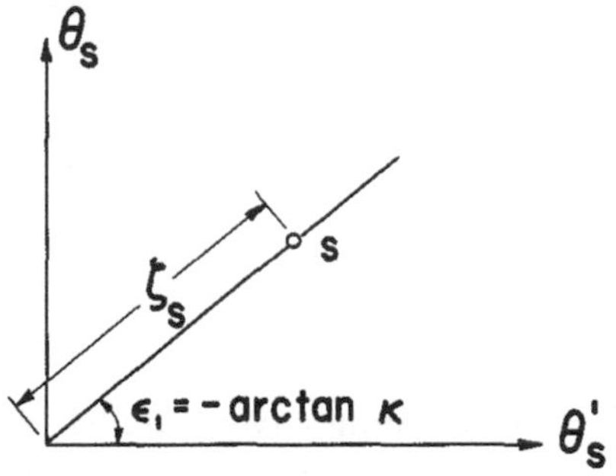

FIG. 96

We have found that $\kappa > 0$ is a necessary condition for a convergent average motion. This condition was to be expected since the same type of control was used here as in the study of the motion of a body with a single degree of freedom. Therefore, we shall study end point regions for $\kappa > 0$ only, these being the regions of practical importance.

If we decide to study only the locations of switch points which might be end points, then we may restrict ourselves to a characterization of these points in a three dimensional space. As coordinates let us choose α_s, v_s/V_0, and $\zeta_s = \sqrt{\theta_s^2 + \theta_s'^2}$ (see Fig. 96). Utilizing the notation of Fig. 96 we have

$$\left.\begin{aligned} \theta_s &= \zeta_s \sin \varepsilon_1 = \zeta_s \frac{-\kappa}{\sqrt{1 + \kappa^2}} \\ \theta_s' &= \zeta_s \cos \varepsilon_1 = \zeta_s \frac{1}{\sqrt{1 + \kappa^2}} \end{aligned}\right\} \tag{206}$$

The condition

$$F' = \theta' + \kappa\theta'' = 0 \tag{207}$$

which determines the boundaries of the end point region, has to be expressed in the coordinates of the new system.

The value of θ'' may be obtained from system L_2 or L_3 corresponding to the type of control used. For system L_2 we obtain from equations (168a),

$$\theta'' = \frac{1}{a_{21}b_{32}} \left\{ (a_{31}a_{20} - a_{21}a_{30})\alpha + \left[a_{31}b_{20} + (a_{31}b_{21} - a_{21}b_{31}) \frac{d}{d\tau} \right] \theta + a_{31}c_{20} \frac{v}{V_0} \pm a_{31}N_2\eta_0 \operatorname{sgn} F \right\} \tag{208}$$

(The lower sign before a_{31} in the last term characterizes system A.) Introducing equation (208) into the expression for F', we get

$$F' = \frac{\zeta_s}{\sqrt{1+\kappa^2}} \left[\left(1 + \frac{a_{31}b_{21} - a_{21}b_{31}}{a_{21}b_{32}} \right) - \kappa^2 \frac{a_{31}b_{20}}{a_{21}b_{32}} \right] + \alpha_s \kappa \frac{a_{31}a_{20} - a_{21}a_{30}}{a_{21}b_{32}} + \frac{v_s}{V_0} \frac{a_{31}c_{20}}{a_{21}b_{32}} \kappa \pm \kappa \frac{a_{31}N_2\eta_0}{a_{21}b_{32}} \operatorname{sgn} F \tag{209}$$

It is clear that $F' = 0$ determines two parallel planes in the "switching space" with the axes ζ_s, α_s, and v_s/V_0. Therefore the boundaries of the end point regions are determined.

It is expected that the region between the two planes will be the end point region. In order to check this idea, we must study the types of switch points and their distribution in the space. In Fig. 97 we see that regular switch points have a certain combination of type, system, and sign of F'. This aids in establishing the following table (see Fig. 98):

Table XI

κ	System and type	F' = slope of F	Sign of F in the interval following the switch point	Portion of "switching space"	
>0	*Aa*	<0	-1	$\left(\frac{v_s}{V_0}\right)_{00} < \left(\frac{v_s}{V_0}\right)_{0011}$	End point for
	Ab	>0	$+1$	$\left(\frac{v_s}{V_0}\right)_{00} > \left(\frac{v_s}{V_0}\right)_{001}$	$\left(\frac{v_s}{V_0}\right)_{0011} < \left(\frac{v_s}{V_0}\right)_{00} < \left(\frac{v_s}{V_0}\right)_{001}$
	Ba	>0	$+1$	$\left(\frac{v_s}{V_0}\right)_{00} > \left(\frac{v_s}{V_0}\right)_{0011}$	Starting point for
	Bb	<0	-1	$\left(\frac{v_s}{V_0}\right)_{00} < \left(\frac{v_s}{V_0}\right)_{001}$	$\left(\frac{v_s}{V_0}\right)_{0011} < \left(\frac{v_s}{V_0}\right)_{00} < \left(\frac{v_s}{V_0}\right)_{001}$

In Fig. 98 the two planes where $F' = 0$ are labelled I and II. Plane I passes through $\zeta_s = 0$, $\alpha_s = 0$, $\left(\frac{v_s}{V_0}\right)_{00\text{I}} = \frac{N_2\eta_0}{c_{20}} > 0$; Plane II passes

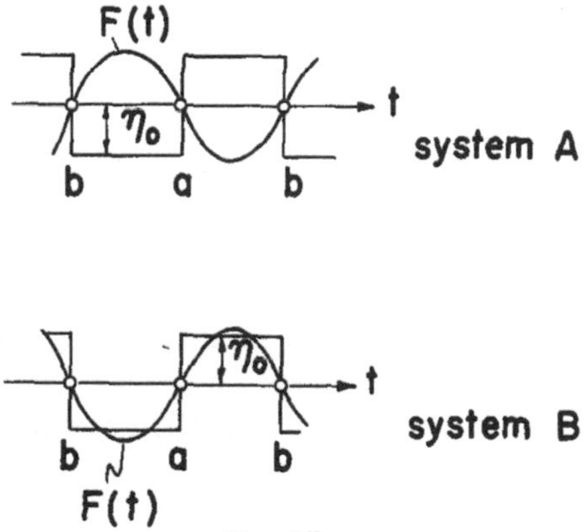

FIG. 97

through $\zeta_s = 0$, $\alpha_s = 0$, $\left(\frac{v_s}{V_0}\right)_{00\text{II}} = -\frac{N_2\eta_0}{c_{20}} < 0$.[1] Portions of the "switching space" created by these planes are characterized by the values of v_s/V_0 for $\zeta_s = 0$ and $\alpha_s = 0$. End point regions are those where none of the combina-

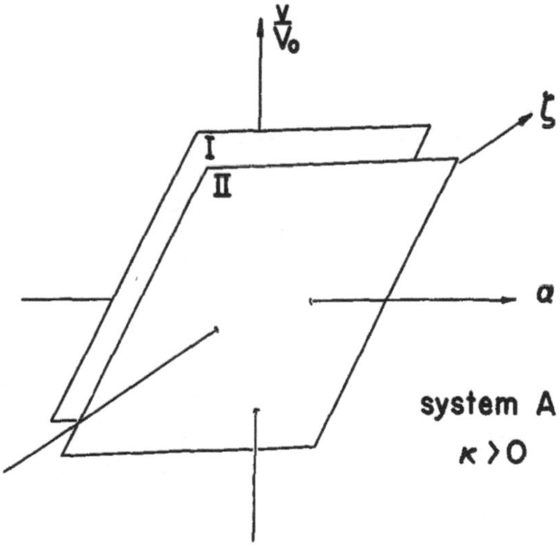

FIG. 98

tions *Aa*, *Ab*, *Ba*, and *Bb* appear. For a given system starting points have both types *a* and *b*. Table XI may easily be extended to include $\kappa < 0$.

[1] Note that $\frac{a_{31}c_{20}}{a_{21}b_{32}} > 0$, $\frac{a_{31}N_2\eta_0}{a_{21}b_{32}} > 0$.

However, it was indicated earlier that we are interested only in the case where κ is positive. The table shows that for $\kappa > 0$ only system A will contain end points. They lie in the space between the two planes I and II, which includes the origin of the coordinate system.

8.6. Periodic motions. The examples have shown that periodic motions occur as asymptotic states of motion. It is of interest to study these periodic motions, which generally represent an undesired asymptotic state. Also, the actual "after-end point" motion tends toward a motion of very small period. It may be recalled that when studying the motion with a single degree of freedom we were able to predict the length of the period as a function of the time lag.

Let us study the behavior of the motion during one period. It will be advisable to start the period at a switch point. The period will consist of two intervals, τ_{p1} and τ_{p2}, each having a different sign of the control function F. Let us call S_i the first switch point, S_1 the switch point after the time τ_{p1}, and S_2 the switch point after the time τ_{p2}. The conditions of periodicity for a continuous phase curve are expressed analytically by

$$\theta_{2e} = \theta_{1i}, \quad \theta'_{2e} = \theta'_{1i}, \quad \alpha_{2e} = \alpha_{1i}, \quad \left(\frac{v}{V_0}\right)_{2e} = \left(\frac{v}{V_0}\right)_{1i} \tag{210}$$

and

$$\theta_{2i} = \theta_{1e}, \quad \theta'_{2i} = \theta'_{1e}, \quad \alpha_{2i} = \alpha_{1e}, \quad \left(\frac{v}{V_0}\right)_{2i} = \left(\frac{v}{V_0}\right)_{1e} \tag{211}$$

Instead of $\alpha_{2e} = \alpha_{1i}$ and $\alpha_{2i} = \alpha_{1e}$, we may choose

$$\left(\frac{v}{V_0}\right)'_{2e} = \left(\frac{v}{V_0}\right)'_{1i} \text{ and } \left(\frac{v}{V_0}\right)'_{2i} = \left(\frac{v}{V_0}\right)'_{1e} \tag{211a}$$

This choice has the advantage that our analytical treatment will be closer to the graphical construction.

If we use the analytical expressions for θ, θ', v/V_0, and $(v/V_0)'$ given in equations (170), the following two systems of equations are obtained (assuming that $(\pm N_2\eta_0 \operatorname{sgn} F)$ is positive in the first interval):

$$\left.\begin{aligned}
\sum_{k=1}^{4} M_{k1} + |B| &= \Sigma M_{k2} e^{\lambda_k \tau_{p2}} - |B| \\
\Sigma \lambda_k M_{k1} &= \Sigma \lambda_k M_{k2} e^{\lambda_k \tau_{p2}} \\
\Sigma C_k M_{k1} - |C| &= \Sigma C_k M_{k2} e^{\lambda_k \tau_{p2}} + |C| \\
\Sigma C_k \lambda_k M_{k1} &= \Sigma C_k \lambda_k M_{k2} e^{\lambda_k \tau_{p2}}
\end{aligned}\right\} \tag{212}$$

and

$$\left.\begin{aligned}\sum_{k=1}^{4} M_{k2} - |B| &= \Sigma M_{k1} e^{\lambda_k \tau_{p1}} + |B| \\ \Sigma \lambda_k M_{k2} &= \Sigma \lambda_k M_{k1} e^{\lambda_k \tau_{p1}} \\ \Sigma C_k M_{k2} + |C| &= \Sigma C_k M_{k1} e^{\lambda_k \tau_{p1}} - |C| \\ \Sigma C_k \lambda_k M_{k2} &= \Sigma C_k \lambda_k M_{k1} e^{\lambda_k \tau_{p1}}\end{aligned}\right\} \tag{213}$$

In addition the condition that $F = 0$ at the switch points gives us

$$\left.\begin{aligned}\Sigma M_{k1}(1 + \kappa\lambda_k) + |B| = 0 \\ \Sigma M_{k2}(1 + \kappa\lambda_k) - |B| = 0\end{aligned}\right\} \tag{214}$$

The system (212) may be rearranged:

$$\left.\begin{aligned}\Sigma(M_{k1} - M_{k2} e^{\lambda_k \tau_{p2}}) &= - 2|B| \\ \Sigma \lambda_k (M_{k1} - M_{k2} e^{\lambda_k \tau_{p2}}) &= 0 \\ \Sigma C_k (M_{k1} - M_{k2} e^{\lambda_k \tau_{p2}}) &= 2|C| \\ \Sigma C_k \lambda_k (M_{k1} - M_{k2} e^{\lambda_k \tau_{p2}}) &= 0\end{aligned}\right\} \tag{215}$$

This offers the possibility of computing the four terms

$$R_{k1} = M_{k1} - M_{k2} e^{\lambda_k \tau_{p2}}$$

as functions of the known values λ_k, C_k, $|B|$, $|C|$. After rearrangement system (213) becomes

$$\left.\begin{aligned}\Sigma(M_{k2} - M_{k1} e^{\lambda_k \tau_{p1}}) &= 2|B| \\ \Sigma \lambda_k (M_{k2} - M_{k1} e^{\lambda_k \tau_{p1}}) &= 0 \\ \Sigma C_k (M_{k2} - M_{k1} e^{\lambda_k \tau_{p1}}) &= -2|C| \\ \Sigma C_k \lambda_k (M_{k2} - M_{k1} e^{\lambda_k \tau_{p1}}) &= 0\end{aligned}\right\} \tag{216}$$

which offers the possibility of computing

$$R_{k2} = M_{k2} - M_{k1} e^{\lambda_k \tau_{p1}}$$

Comparison of the two rearranged systems shows that

$$M_{k1} - M_{k2} e^{\lambda_k \tau_{p2}} = -(M_{k2} - M_{k1} e^{\lambda_k \tau_{p1}}) \tag{217}$$

or

$$M_{k1}(1 - e^{\lambda_k \tau_{p1}}) = - M_{k2}(1 - e^{\lambda_k \tau_{p2}})$$

Introducing equation (217) into equation (214) gives

$$\left.\begin{aligned}\Sigma M_{k1}(1 + \kappa\lambda_k) + |B| &= 0 \\ -\Sigma M_{k1} \frac{1 - e^{\lambda_k \tau_{p1}}}{1 - e^{\lambda_k \tau_{p2}}}(1 + \kappa\lambda_k) - |B| &= 0\end{aligned}\right\} \tag{218}$$

These two equations are compatible only if

$$\tau_{p1} = \tau_{p2} \tag{219}$$

Thus we have obtained the result that the period is composed of two intervals of equal length. With $\tau_{p1} = \tau_{p2}$ we obtain, from equation (217)

$$M_{k1} = -M_{k2} = M_{kp} \tag{220}$$

Either system (212) or (213) may be used to compute for a chosen length of half period τ_p the value of M_{kp}. Let us choose (212), or the rearranged form (215), which gives

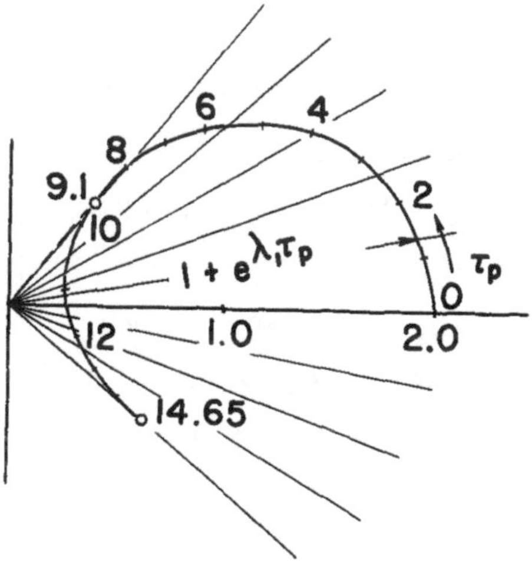

FIG. 99

$$\left.\begin{aligned} \Sigma M_{kp}(1 + e^{\lambda_k\tau_p}) &= -2\,|B| \\ \Sigma M_{kp}\lambda_k(1 + e^{\lambda_k\tau_p}) &= 0 \\ \Sigma C_k M_{kp}(1 + e^{\lambda_k\tau_p}) &= 2\,|C| \\ \Sigma C_k\lambda_k M_{kp}(1 + e^{\lambda_k\tau_p}) &= 0 \end{aligned}\right\} \tag{221}$$

This system of linear equations for the four terms $M_{kp}(1 + e^{\lambda_k\tau_p})$ has the same coefficients as does the system of equations for the ξ_k. Thus the solution is easily obtained in the form

$$M_{kp}(1 + e^{\lambda_k\tau_p}) = 2\,|C|\,\mathfrak{E}_{1k} - 2\,|B|\,\mathfrak{E}_{3k} \tag{222a}$$

For $\tau_p = 0$: $M_{kp} = 2\,|C|\,\mathfrak{E}_{1k} - 2\,|B|\,\mathfrak{E}_{3k}$

As these are constants for a given missile, they may be used conveniently for studying the dependency of M_{kp} on τ_p. Thus we may write

$$M_{kp} = M_{kp}(0) \cdot \frac{1}{1 + e^{\lambda_k\tau_p}} \tag{222b}$$

For the first root λ_1 the denominator $1 + e^{\lambda_1\tau_p}$ is plotted as a function of τ_p in Fig. 99. This provides a survey of the values of M_{1p} which may occur.

λ_3 has a large negative real value; therefore $1 + e^{\lambda_3 \tau_p}$ tends rapidly towards unity. It is not plotted.

M_{kp} would have had the opposite sign if we had assumed $[\pm N_2 \eta_0 \operatorname{sgn} F]$ to have been negative in the first interval. M_{kp} determines immediately s_1 and s_3; the terms $\lambda_k M_{kp}$ will help us compute s_1^* and s_3^*. Thus the coordinates of switch points of the periodic motion in the partial phase planes are known. By connecting the switch points with the appropriate spirals in the s_1, s_1^* and s_3, s_3^* planes we can easily see the system to which the motion belongs.

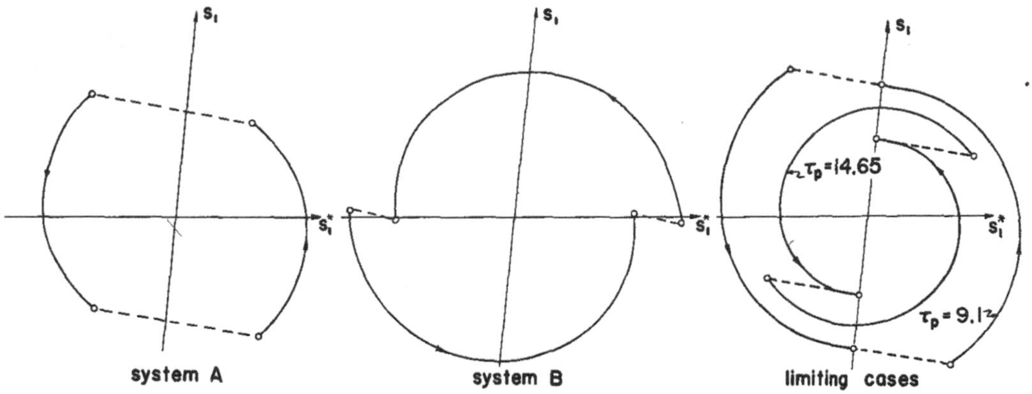

FIG. 100a FIG. 100b FIG. 100c

Fig. 100 indicates that there are two characteristic forms of partial phase curves in the s_1, s_1^* plane, one belonging to system A, the other to system B. Fig. 100c shows the transitory case between periodic motion of system A and system B; it occurs for $\tau_p = 9.1$ and $\kappa = \infty$. Furthermore, it is obvious that there is an upper limit to τ_p, given by the fact that the jumps Δs_k and Δs_k^* (independent of τ_p) must form a closed curve when joined to the two spiral segments. Fig. 100c shows this limiting case, for which $\tau_p = 14.65$ and $\kappa = \infty$. After exceeding this limit, only rest of control appears, if one starts with a point chosen from equation (222).

Having computed M_{kp}, the values of θ_p, θ_p', α_p, and $(v/V_0)_p$ at the switch points of the periodic motion may be obtained

$$\left.\begin{aligned} \theta_p &= \Sigma M_{kp} + B \\ \theta_p' &= \Sigma \lambda_k M_{kp} \\ \left(\frac{v}{V_0}\right)_p &= \Sigma C_k M_{kp} + C \\ \alpha_p &= \Sigma A_k M_{kp} + A \end{aligned}\right\} \tag{223}$$

8.7. Discussion of the general behavior of a missile under discontinuous position control. After having studied in detail some examples of the motion of a controlled missile, it is possible to summarize the different types of motion which may occur and to state the common traits and differences between controlled motion with one or more degrees of freedom. For ideal control (without imperfection) there are three final states of motion; periodic motion, motion leading to an end point, and motion under

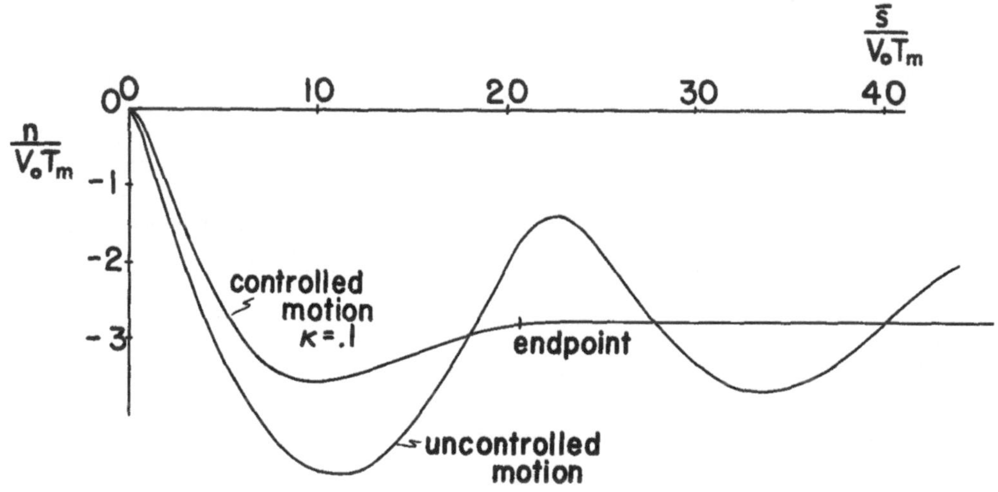

FIG. 101

rest of control. The mathematical proof that periodic motion is the asymptotic state of any motion not tending toward an end point or toward a motion with control at rest is still missing. However, experience indicates the fact. For a periodic motion the energies put in and withdrawn by the control mechanism during a period are equal in magnitude.

Generally the control mechanism is used to keep the missile as close as possible to the desired path. This means that the mechanism should be so designed as to reduce a disturbance by withdrawing energy more rapidly than would be the case for an uncontrolled but stable missile. Hence the periodic motion with an average change of energy equal to zero is undesirable. If the periodic state is an asymptotic one, reached after considerable energy withdrawal, and if it has sufficiently small amplitude, the situation is different. This state will appear in the so-called "after-end point" motion with imperfect control systems. However, here is an essential difference compared with systems with one degree of freedom. There the "after-end point" motion is entirely determined by the coefficient κ and the delay of switching. This is not the case for more degrees of freedom. The coefficients of the equation of motion play a decisive role, as is shown on pp. 149-151.

The discussions in the preceding chapter have shown that for a successful damping of disturbances system A and a small positive value of the coefficient κ should be chosen. The size of the control angle η_0 (or η_{h0}) has to be chosen so that all expected disturbances produce motions leading rapidly to an end point for ideal control (see Table XI).

In Fig. 101 the deviation from a desired path for a disturbed missile with a control mechanism is represented. The effect of control shows a strong damping; however, a finite (but constant) deviation will remain. This is due to the type of control ($F = \theta + \kappa\theta'$) which has been chosen. In order to return to $n = 0$ for $t \to \infty$, one has to choose a control which checks the amount $n = V_0 T_m \int_0^T (\theta - \alpha) d\tau$ at every instant and employs it for correction.

We have studied only one type of control function. In a system with three degrees of freedom, described by four phase variables, there exist quite a few possibilities for control. Corresponding to known continuous control systems, different types of discontinuously working systems may be designed. In general they will be divided into two groups: those which have discontinuous derivatives of the control function, and those which do not. As an example of the latter we may look at

$$F = a\frac{v}{V_0} + b\theta$$

which is sometimes suggested as a control function. Then

$$F' = a\left(\frac{v}{V_0}\right)' + b\theta'$$

will be a continuous function, even in the switch points, as shown in Table X. Hence we cannot expect end points and the advantageous "after-end point" motion of imperfect control systems to appear. To obtain this advantage we would have to have

$$F = a\left(\frac{v}{V_0}\right) + b\theta + c\theta'$$

Thus the knowledge gained in studying in detail the motion of a missile under a special control system offers the possibility of judging other control systems without too much difficulty.

APPENDIX I

Table VII. Notation

c = chord

b = span

S = wing area

W = weight

m = mass

ρ = density

V_0 = velocity of the undisturbed missile

t = time

$T_m = \dfrac{2m}{\rho S V_0}$ = special time unit

$\tau = \dfrac{t}{T_m}$ = nondimensional time variable

$\bar{\mu} = \dfrac{2m}{\rho S c}$

$I_y = mk_y^2$ = moment of inertia about y axis

$q_0 = \rho V_0^2/2$ = dynamic pressure

T = thrust

x, y, z = rectangular coordinates (wind axes)

α = angle of attack

θ = angle of pitch

γ = flight path angle $[\gamma = \theta - \alpha]$

$\varepsilon = k_1\alpha$ = downwash angle at the tailplane

v = increment of velocity in x direction

$\bar{s}$ = coordinate along the path of the undisturbed flying missile

n = normal deviation from undisturbed path of the missile

i_w = wing incidence angle (relative to thrust line)

η = angle of aileron

η_h = angle of elevator

$N_2 = \dfrac{\partial L}{\partial \eta}\dfrac{1}{q_0 S}$ = rate of change of lift with η

$N_3 = \dfrac{\partial M}{\partial \eta_h}\dfrac{1}{q_0 S c}$ = rate of change of moment with η_h

The subscript 0 refers to undisturbed straight flight of the missile.

$$C_L = \frac{\text{lift}}{q_0 S}$$

$$C_D = \frac{\text{drag}}{q_0 S}$$

$$C_M = \frac{\text{pitching moment}}{q_0 S c}$$

$$C_W = \frac{W}{q_0 S}$$

$$C_{D_0}^* = C_{D_0} - \left(\frac{\partial T}{\partial q_0}\right)_{\alpha_0} \frac{\cos(\alpha_0 - i_w)}{S}$$

$$C_{L_0}^* = C_{L_0} + \left(\frac{\partial T}{\partial q_0}\right)_{\alpha_0} \frac{\sin(\alpha_0 - i_w)}{S}$$

$$C_{D_0}' = \left(\frac{dC_D}{d\alpha}\right)_{\alpha_0}$$

$$C'_{L_0} = \left(\frac{dC_L}{d\alpha}\right)_{\alpha_0} \qquad \dot{\alpha} = \frac{d\alpha}{dt}$$

$$C'_{M_0} = \left(\frac{dC_M}{d\alpha}\right)_{\alpha_0} \qquad C_{Mq} = \left(\frac{dC_M}{dq}\right)_{\alpha_0}$$

$$q = \frac{d\theta}{dt} = \dot{\theta} \qquad C_{M\dot{\alpha}} = \frac{d\varepsilon}{d\alpha} C_{Mq} = k_1 C_{Mq}$$

Table VIII. Numerical Values for Disturbance of the Gliding Flight

$$C_{L_0} = 0.213$$
$$C_{D_0} = 0.022$$
$$C'_{L_0} = 3.55$$
$$C'_{M_0} = -0.365$$
$$C'_{D_0} = 0.197$$

$$C_{Mq} = -10.0$$
$$\bar{\mu} = 550$$
$$T_m = 3.25 \text{ sec.}$$
$$\frac{d\varepsilon}{d\alpha} = k_1 = 1.6$$
$$\frac{k_y^2}{c^2} = 1.66$$

$$a_{10} = C'_{D_0} - C_{L_0} = -0.016$$
$$a_{20} = C'_{L_0} + C_{D_0} = 3.57$$
$$a_{21} = 1$$
$$a_{30} = \bar{\mu} C'_{M_0} = -201$$
$$a_{31} = C_{M\dot{\alpha}} = -16$$

$$b_{10} = C_W \cos \gamma_0 = 0.213$$
$$b_{20} = C_W \sin \gamma_0 = -0.022$$
$$b_{21} = -1$$
$$b_{31} = C_{Mq} = -10$$
$$b_{32} = -\frac{k_y^2}{c^2} = -1.66$$

$$c_{10} = 2C^*_{D_0} = 0.044$$
$$c_{11} = 1$$
$$c_{20} = 2C^*_{L_0} = 0.426$$

$$N_2 \eta_0 | = 0.05$$
$$\lambda_{1,2} = -0.02937 \pm 0.2777i$$
$$\lambda_{3,4} = -9.609 \pm 7.062i$$

	$k = 1$	$k = 3$
A_k	$0.01383e^{4.842i}$	$1.282e^{0.229i}$
C_k	$0.7658e^{1.625i}$	$0.01624e^{0.612i}$
λ_k	$0.2792e^{1.676i}$	$11.93e^{2.508i}$
$\mathfrak{E}_{1k}$	$0.6533e^{4.697i}$	$0.01571e^{0.880i}$
$\mathfrak{E}_{2k}$	$0.3054e^{3.054i}$	$0.5135e^{5.347i}$
$\mathfrak{E}_{3k}$	$0.5012e^{6.230i}$	$0.0006072e^{3.746i}$
$\mathfrak{E}_{4k}$	$0.01256e^{6.264i}$	$0.05530e^{4.483i}$

APPENDIX II

CONSTRUCTION OF LOGARITHMIC SPIRALS BY AN APPROXIMATE METHOD[1]

A logarithmic spiral of the type shown in Fig. 102 may be represented by the relation

$$r = r_0 e^{-m\theta}$$

where, corresponding to the applications of this report,

$$m = \cot \alpha = \frac{D}{\nu} \text{ and } \theta = \nu\tau$$

The method of construction suggested here has been found to give good results for values of D as high as 0.5. For values greater than this the method should be employed with discretion.

Assume that the points P_0 and P_1 (Fig. 102) are given as exact points on the spiral for $\theta = 0$ and $\theta = \pi/2$. Using the radii r_0 and r_1 of the two points P_0 and P_1 one must now compute the value d_0 from the relation

$$d_0 = \frac{r_0 - r_1}{\sqrt{2}}$$

Now construct a line Oa at 45° to the positive x axis and mark off the distance d_0 along this line, measuring from the origin, thus fixing the point N_0. Then (referring to Fig. 102) it may be seen that $\overline{P_0N_0} = \overline{P_1N_0}$ since the triangles P_1LN_0 and P_0ON_0 are congruent. Now it can be shown that a circle, with N_0 as center and radius equal to $\overline{P_0N_0} = \overline{P_1N_0} = \rho_m$, will pass through a third exact point $P_{45°}$, of the spiral:

$$\rho_m^2 = \left(r_0 - \frac{d_0}{\sqrt{2}}\right)^2 + \left(\frac{d_0}{\sqrt{2}}\right)^2 = \left(\frac{r_0 + r_1}{2}\right)^2 + \left(\frac{r_0 - r_1}{2}\right)^2$$

$$(\overline{P_{45°}O})^2 = \rho_m^2 - d_0^2 = r_0 r_1$$

Since

$$r_1 = r_0 e^{-\cot \alpha(\pi/2)}$$

[1] This method is due to Professor L. S. Jacobsen, Stanford University, and appreciation is expressed to him for allowing the use of this unpublished material.

it is found that

$$(\overline{P_{45°}O})^2 = r_0^2 e^{-\cot\alpha(\pi/2)} = r_{45°}^2$$

It is now clear that the circle given by a radius of ρ_m and center N_0 will pass through the exact points of the spiral P_0, P_1, and $P_{45°}$. The deviations

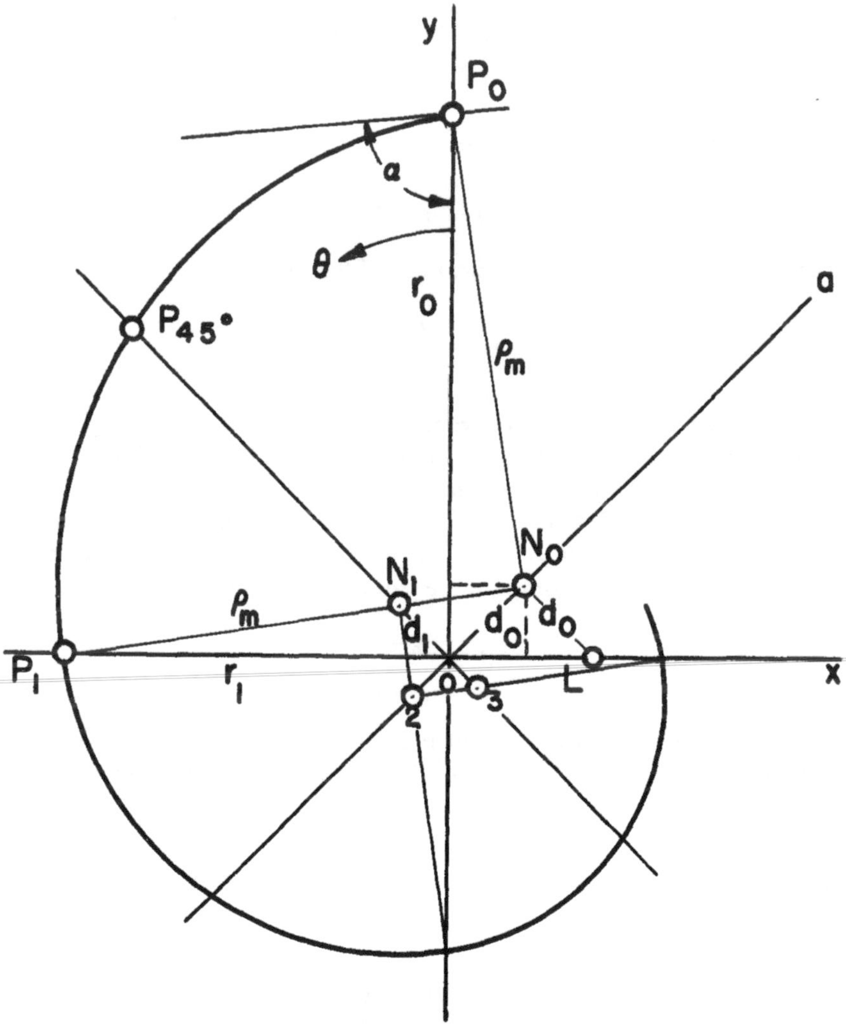

FIG. 102

of this circular arc from the corresponding portion of the spiral will be so small as to be undetectable in a finished drawing, so long as the restriction imposed on D mentioned earlier is not disregarded.

If one would desire to continue the construction of the spiral past this stage, it would appear necessary to repeat the procedure just given for the region between $\theta = \pi/2$ and $\theta = \pi$. The result of such a construction would be to have the center of the circular arc at the point N_1, a distance d_1 from the origin measured along a 45° line in the second quadrant. Then we would have

$$d_1\sqrt{2} = r_1 - r_2 = r_1\,(1 - e^{-\cot\alpha(\pi/2)})$$

but, since we had previously obtained

$$d_0\sqrt{2} = r_0 - r_1 = r_0(1 - e^{-\cot\alpha(\pi/2)})$$

we see that

$$\frac{d_0}{d_1} = \frac{r_0}{r_1}$$

and, therefore, N_1 lies on the line $\overline{P_1N_0}$ since the triangles P_0ON_0 and P_1ON_1 are similar (also $\measuredangle N_1OP_1 = \measuredangle N_0OP_0 = 45°$). This analysis shows that constructions succeeding the original one may be made without computation of the d_n values. More simply, if lines at 45° to the axes are drawn in each of the other quadrants, the center for the desired circular arc is readily found as the point where the last radius of the preceding arc intersects the 45° line of its quadrant.

INDEX

Milton Keynes UK
Ingram Content Group UK Ltd.
UKHW011311041123
431920UK00005B/119

9 780691 627182